D0437891

# IN SEARCH OF
# THE
# LOST
# CORD

## Solving the
## Mystery of
## Spinal Cord Regeneration

# LUBA VIKHANSKI

Joseph Henry Press
Washington, D.C.

Joseph Henry Press • 2101 Constitution Avenue, N.W. • Washington, D.C. 20418

The Joseph Henry Press, an imprint of the National Academy Press, was created with the goal of making books on science, technology, and health more widely available to professionals and the public. Joseph Henry was one of the founders of the National Academy of Sciences and a leader in early American science.

The Dana Press, a division of The Charles A. Dana Foundation, publishes health and popular science books about the brain for the general reader. The Dana Foundation is a private philanthropic organization with particular interests in health and education.

Any opinions, findings, conclusions, or recommendations expressed in this volume are those of the author and do not necessarily reflect the views of the National Academy of Sciences or its affiliated institutions.

**Library of Congress Cataloging-in-Publication Data**

Vikhanski, Luba.
  In search of the lost cord : solving the mystery of spinal cord regeneration / Luba Vikhanski.
    p. ; cm.
  Includes bibliographical references and index.
  ISBN 0-309-07437-1 (alk. paper)
  1. Spinal cord—Regeneration. 2. Spinal cord—Surgery.
  [DNLM: 1. Spinal Cord Injuries—rehabilitation. 2. Nerve Regeneration. 3. Research. WL 400 V694i 2001] I. Title.
  RD594.3 V553 2001
  617.4'82—dc21
                                                                2001039846

Printed in the United States of America

*To the memory of my father*

# Contents

## Part I
## The Science of the Impossible

# Part II
# The Many Faces of Hope

# Author's Note

When Dr. Wise Young, a veteran spinal cord researcher, suggested that I interview about 100 scientists for this book, I hoped I hadn't heard him correctly. My idea was to cover a few highlights in spinal cord regeneration and repair; this, I thought, could be accomplished by describing the work of five or six key laboratories. However, I soon became fascinated by the recent reversal of attitude, toward optimism and hope, in this field of research. Trying to grasp the scope of this reversal, I attended four scientific conferences and interviewed some 150 people in eight countries, most of them scientists and students working on the spinal cord and on the related topic of brain repair. Their names appear in the Special Acknowledgment section at the end of this book.

I am deeply grateful to all the researchers for allowing me to interview them at length. Describing their work even briefly would fill many volumes, and I had to make some difficult choices. This book is not by any means an all-encompassing history of spinal cord research. Rather, I chose several landmark research efforts that to me told the story of spinal cord regeneration and repair. I tried to highlight scientists who created a new line of research and stayed with it, but I'm sure I left out many important contributions. I am solely responsible for the selection of projects described in this book, and I apologize to all the researchers whose work I did not include.

Initially, the Dana Press suggested I write a book on spinal cord regeneration together with Professor Michal Schwartz, but when Schwartz was unable to find the time for the project, I was given the opportunity to write the book on my own. I am grateful to Professor Schwartz for her comments and support throughout my work.

Professor Young, whom I first met when this book was still an idea, gave me an inspiring and informative introduction to the field. Without knowing it, he continued to help me for an entire year. While working on the book, I regularly consulted the web site, now at *www.SCIWire.org*, where he posts popular scientific articles and answers questions about spinal cord research. The entries posted on the site by people with spinal cord injury allowed me to appreciate the urgency that the quest for spinal cord repair has in their lives.

Numerous scientists spent a long time talking with me about the field in general and explaining basic concepts. I would like in particular to acknowledge the help of Professors Albert Aguayo, Bernice Grafstein, Marion Murray, Lars Olson, Constantino Sotelo, and John Steeves. I am grateful to all the scientists who reviewed drafts of chapters dealing with their own work or with research in related fields, and to Drs. Jerald Bernstein, Naomi Kleitman, and Michael Rasminsky for commenting on parts of the manuscript.

Without the assistance of people at different institutions, the job of conducting research for this book would have been nearly impossible. At the Cajal Institute, in Madrid, I benefited from exceptional hospitality extended to me by Professor Javier DeFelipe and by members of his laboratory. I would also like to express my thanks to the staff of the Institute's library, to Dr. Miguel Freire, responsible for the Institute's Cajal Legacy, to Maria A. Pérez-de-Tudela, to members of the Tello family, and to Maria Angeles Ramón y Cajal for the opportunity to visit Cajal's last residence.

At the University of California, Los Angeles, home to Dr. William Windle's papers, I was welcomed and assisted by Dr. Louise Marshall, director of the Neuroscience History Archives, and benefited from the professional help of Teresa Johnson and other staff members of the history division in the Louise M. Darling Biomedical Library.

In other countries, my special thanks go to Ida Engqvist at the Karolinska Institute in Stockholm, to Silvia Kaufmann at the University of Zurich, to Doris Soares at Montreal General Hospital, and to Victor Swoboda at Montreal Neurological Institute.

Sam Maddox provided me with journalistic insights and tips based on his own experience in writing about spinal cord repair. His book, *The Quest for Cure: Restoring Function After Spinal Cord Injury*, was the launching pad for my work and an invaluable resource.

Friends and colleagues generously took the time to read parts of the manuscript at different stages and made indispensable comments. They are Leora Frankel, Riza Jungreis, Joe Katz, Anna Kiselev, and Rinna Samuel. In addition, I am grateful to Sandra Cash and to Hugh Aldersey-Williams for editorial comments and to Efrat Regev for help with computers.

Since this book is a coproduction of two publishers, I had the benefit of support from two editorial teams. It was a pleasure and honor to work with Jane Nevins and Andrew Cocke from the Dana Press and with Stephen Mautner from the Joseph Henry Press. These editors put so much time and thought into the manuscript that on more than one occasion I felt the book was truly a team effort.

This book is the brainchild of Jane Nevins, editor in chief of the Dana Press, who came up with its idea and inspired my interest in the topic. Throughout my work, she provided me with the kind of help and support that I would have expected to receive from a co-author or a friend. Without her enthusiasm, this book would not have been started—and would never have been finished.

Luba Vikhanski
Tel Aviv, Israel
2001

# Introduction

"**A** disease that cannot be treated" was how an Egyptian papyrus referred to spinal cord injury more than 3,500 years ago. When this first medical document to describe spinal cord trauma was translated into English in 1930, the verdict was picked up by medical textbooks around the world as still sadly relevant: Several millennia had passed, but physicians were still unable to treat injuries to the spinal cord. Only in the middle of the twentieth century did medicine learn to keep spinally injured people alive. The resulting paralysis, however, was considered irreversible. Dead nerve cells in the spinal cord, scientists believed, could not be replaced; severed nerve fibers would never regenerate; paralyzed people would never walk again. With hardly any research in this area, there was no progress, which in turn reinforced the impression that the problem was intractable.

But in the past two decades, the attitude toward spinal cord injury changed dramatically. In the 1980s, thanks to new discoveries and tools, research in the field of spinal cord regeneration and repair began to gain credibility and momentum. The enormity of the challenge—repairing a structure once believed to be irreparable—created an appeal some researchers could not resist. It offered an opportunity to leave a mark, to make headway in uncharted waters. At about the same time, the entire thinking about the central nervous system, consisting of the brain and spinal cord, underwent a revolutionary transformation. Once considered a fixed, rigid network of nerves, the system is now viewed as flexible, malleable—and hopefully, reparable. There is now hope that one day, people with paralysis will be able to regain at least some of their lost abilities: breathing, sexual function, bladder and bowel control, walking. Implications for other diseases of the nerves, particularly such brain disorders as stroke and Alzheimer's and Parkinson's diseases, can also be great, as spinal cord and brain research are closely intertwined.

The new hope touches the lives of millions of people. In 2001, individuals paralyzed by trauma to the spinal cord numbered 250,000 in the United States alone and an estimated 2 million worldwide. Leading causes vary in different parts of the world: traffic accidents in the United States, gunshot wounds in Brazil, earthquakes and coal mine accidents in China, and falls from coconut trees in the Philippines. Some population groups are more vulnerable than others; in many countries, young men, the group most prone to risk-taking behavior, account for more than half of the cases. But anyone anywhere is a potential victim: Some 85,000 new spinal cord injuries around the world—10,000 of them in the United States—put people in wheelchairs every year. These illnesses annually affect more than 50 million Americans and cost society more than $400 billion, according to the Washington-based Society for Neuroscience.

Today hundreds of scientists in different countries are striving to regenerate and repair damaged nerves in the spinal cord, and the pace of their research continues to accelerate. "Now it's quite accepted that regeneration is a serious scientific field that is going somewhere, not a bunch of lunatics pursuing some crazy notion," says Dr. Fredrick Seil, who for 15 years organized annual regeneration conferences on behalf of the Department of Veterans Affairs in the United States. But as recently as the mid-1990s, sanctioned human trials for most therapies were only a distant prospect. Then, toward the end of the decade, reports of experiments in humans began to trickle in. Fetal cell trials were launched in Sweden and the United States. In 1999, an Israeli trial got a green light from the U.S. Food and Drug Administration (FDA). In 2000, two other trials, to be conducted in the United States, were approved by the FDA. By 2001, the regeneration field was abuzz with talk about sanctioned human experiments expected to begin in the first decade of the twenty-first century. An FDA approval of a trial does not mean that a therapy is safe and effective—these are precisely the unknowns the trial is designed to resolve. Nor is a single therapy likely to provide a definitive cure for paralysis; scientists believe that if the task is doable, it will take a combination of different approaches to fully repair a complex organ like the spinal cord. The exciting part, however, is that many such approaches are in the works. Will these approaches work in humans? As several regeneration therapies reach the finish line, the answers are nearer, the suspense mounts.

Spinal cord repair involves many goals, but this book focuses mainly on regeneration, a compelling quest to reverse the irreversible, to undo a fatal mistake. It does, however, refer to other aspects of repair, such as "reeducating" the spinal cord after injury, as well as to certain brain research projects, while the appendix offers a review of spinal cord anatomy and several basic concepts of neuroscience. This book is the tale of the turnaround that has occurred in spinal cord research and of the people who make hope real—their failures and successes and their dedication to a scientific pursuit of a goal once considered an impossible dream.

# IN SEARCH OF
# THE
# LOST
# CORD

# Prologue

On the late afternoon of Sunday, June 25, 2000, Roy Holley was driving to his home in Ridgway—a tiny old cowboy town tucked away at 7,000 feet on one of western Colorado's most scenic roads snaking through the Rocky Mountains. Holley was getting ready for a busy night, preparing the training seminar for executives he was to conduct on Monday, but his plans were abruptly cut off by the sight that opened in front of his eyes on a relatively straight stretch of the main highway. There, lying crumpled in a ditch by the side of the road, was the dark-green Toyota Corolla belonging to his teenage daughter Melissa; its hulk was destroyed. "My first thought was that Melissa was dead; looking at the car I didn't see how anybody could have lived through that accident," he says. The crash had occurred barely an hour earlier; a policeman examining the scene told him Melissa had survived and been taken to the hospital.

Melissa, who had come to spend the summer with her parents

in Ridgway, had just finished her freshman year at Harding University, a Christian college in Arkansas. A tall, pretty young woman with long, honey-colored hair, high cheekbones, and her father's dreamy green eyes, she had been featured in the *Who's Who* of American high schools as an outstanding student, the captain of her high school's women's soccer team, an avid swimmer, and a scuba diver. She had been planning to spend the fall semester in Australia through her college; in the meantime she had gotten two waitressing jobs in the Ridgway area to make money for the trip. Around 4 p.m. on June 25, as she was driving from one job to another, it started to rain. She slid off a curve, the Toyota did three flips; Melissa flew through the shattered windshield into the ditch while the car continued to flip, finally landing over her body. "When I woke up it was on top of me, the wheel was right here by my side, and I didn't remember the accident. I suppose I had a blackout, so I was just, like, where am I?" she recalled. Miraculously, Melissa suffered hardly any injuries to other parts of her body, but the car, which needed to be cut apart in order to free her, had crushed two vertebrae, T6 and T7, located in the midchest area of her spinal cord. The injury had disconnected the vital nerve pathways between her brain and lower body, leaving her paralyzed from the chest down.

Spinal cord injuries often have disastrous consequences: The cord, almost an inch thick, is responsible for all communication between the brain and the rest of the body. Information flows both ways. In one direction, the motor nerve pathways responsible for movement carry nerve impulses from the brain down the spinal cord, where messages are relayed to various parts of the body. In the other, sensory pathways carry impulses upward from the skin and internal organs to the relevant segment of the spinal cord, and from there to the brain.

Looking at the spinal cord, no one would suspect that it contains all this complicated circuitry. The cord's jellylike outer part consists of so-called white matter. Only under the microscope is it possible to see that the milky, smooth white matter is made up of countless fibers, the cablelike extensions of nerve cells. Some, arising from the brain cells, extend to great distances; other fibers come from nerve cells that sit in the inner part of the spinal cord, the soft, mushy tissue known as gray matter. The gray matter core is where all the action takes place: Here information is processed and integrated; the white matter on the outside is the highway along which signals travel.

When the cord is damaged, the muscles below the injury level do not know what to do without commands from the brain; the brain, in its turn, loses touch with the body below the injury. The higher the injury, the more life-threatening it is likely to be. The topmost vertebrae surround the spinal nerves that control the neck, shoulders, and upper arms as well as breathing and the diaphragm, so that an injury to the neck area can, and frequently does, interfere with the person's ability to breathe. Arguably the most famous spinally injured person in the world, movie star "Superman" Christopher Reeve, who broke his neck in a fall from a horse in 1995, has just about the highest possible injury. It affects the nerves within the two vertebrae closest to the brain, C1 and C2, and is known as the "hangman's injury" because that is where the neck breaks when the noose tightens. Immediately after his accident, Reeve could not breathe without a ventilator and was able to move only his head; that movement was thanks to the head-turning muscles controlled by nerves that travel outside the spinal cord, directly from the brain to the muscles. Paralysis from the neck area down is called "quadriplegia" or "tetraplegia," referring to the loss of movement in all four limbs. People with midback injuries can still move their upper torsos, their arms, and their hands; the outcome of such injuries is referred to as "paraplegia," a term that is sometimes used to describe all types of paralysis.

In addition to the level of injury, the consequences of spinal cord damage depend on whether an injury is "complete" or "incomplete." The spinal cord is rarely actually cut in two—this tends to happen only as a result of gunshot or stab wounds; most often, the cord is bruised or crushed. Still, an injury is described as "complete" if the person has no movement or feeling in all areas of the body below the point of injury. Even insignificant recovery of function after such injuries is rare. In contrast, people with "incomplete," or partial, injuries sometimes eventually regain substantial body function.

"Getting out of the wheelchair" is a common metaphor for a cure for spinal cord injuries, but paralysis does much more than prevent walking. Injured people can lose voluntary control of their bowel and bladder activity, their sexual function may be impaired, and they may lack sensation in the paralyzed part of the body. For people with high-cord injuries, even the simplest activities, such as holding a spoon or typing on a computer, may be out of reach. "On the so-so days in this wheelchair, I would sell or give away whatever I have to be able to use my arms and hands fully; on the bad days I think I'd sell my soul," a spinally injured man from Eugene, Oregon, who called himself X-Racer wrote in an internet discussion forum.

One of the more devastating consequences of such severe injuries is dependence on other people. "I don't really *need* to walk again, but I do need, and desperately, desperately want my independence back," says Barbara Turnbull, a Canadian journalist paralyzed from the neck down since she was 18, when a bullet passed through her spine during an armed robbery at the convenience store where she worked. Turnbull, now in her 30s, holds a full-time job and travels around the world, but because the only body part she can move is her head, she relies heavily on outside help. "I want to get myself out of bed without assistance whenever I want, and I want to go to bed, the same way, whenever I want," Turnbull writes in her autobiography. "I want to read again before I go to sleep at night, and I want to reach over with my arm and snap off the light with my hand. I want to lounge on the couch on a rainy weekend with a good book. I want to get in a car and drive again, by myself. And I believe now, with all my heart and mind, that it will happen."

## History in the Making

At the hospital in Grand Junction, the nearby city to which Melissa Holley was flown after her accident, the neurosurgeon told her she would never walk again. But Roy Holley wouldn't accept the verdict. At once, he turned to the internet and, within 48 hours, he had learned about a new spinal cord regeneration therapy developed in Israel and about to be tested at the Sheba Medical Center near Tel Aviv. The Israeli researchers had already received approval for their trial from the U.S. Food and Drug Administration and were waiting for the first patient. Holley sent a fax to the hospital, not really expecting an answer. "You know, you fax something from Grand Junction, Colorado, to a general fax number of a hospital in Israel, and you don't even know if it'll ever get to the right person. But amazingly enough, six hours later, I got a response from Dr. Valentin Fulga at Proneuron, the biotech company running the Israeli trial." Now Melissa had to undergo the experimental procedure no later than two weeks from the time of the accident, and she had to meet a long list of criteria for the trial.

Luckily, Roy Holley and his wife, Gwen, own a small company that provides training aimed at honing the leadership skills of executives. In fact, they had chosen to settle in Ridgway with their two younger sons because the serene mountain resort provided a perfect setting for their leadership seminars. Flying their critically injured daughter halfway around the globe in time to receive the Israeli treatment was to prove for them a true exercise in decisiveness and leadership. Melissa was still too ill to take a regular flight, but within two days the Holleys had organized a $90,000 loan to hire a private ambulance jet, created a Melissa Holley Medical Fund at the First National Bank in Ridgway, and told their story to the local media in the hope that, following newspaper and television reports, people from all over the United States would donate money to the fund, as indeed they did.

On July 3, Melissa and Roy embarked on a 17-hour flight to Tel Aviv in a small private jet that operated like a one-patient intensive care unit, with a doctor and a nurse aboard. Melissa was still hooked to an intravenous infusion; her paralyzed body was enclosed in a large plaster brace and strapped to a gurney to stabilize her spinal cord. On July 9, two days after her nineteenth birthday, she became the first person to undergo the brand-new Israeli procedure at the

Sheba Medical Center. A neurosurgeon opened the membranes of her spinal cord and injected into the injury site several million immune cells removed from her own blood, whose purpose was to stimulate her damaged spinal cord fibers to regrow. During the operation, the surgeon also placed two titanium rods into Melissa's back to stabilize and straighten the spinal column distorted by the weight of the car.

At Sheba, a sprawling hospital complex enlivened by bright subtropical vegetation, I met Melissa in late August of 2000, while she was undergoing rehabilitation and coping with the shock of being so unexpectedly transplanted into a foreign environment. I had already seen countless images of nerve fibers regenerating in rat spinal cords and now, for the first time, I myself was meeting a human being who might benefit from the years of animal research. Roy Holley, a big man who speaks in a surprisingly soft, reserved manner, told me that it was Melissa herself who had made the ultimate decision to enroll in the experiment: "I had no concerns once I learned the treatment was approved by the FDA. Then Dr. Fulga was incredibly good at providing us with extensive information; I felt I could trust him. And it's not like there were any other options. It was really, try this therapy or just accept the fact Melissa wouldn't walk again. And that would have been the harder thing to do."

In the first two months after the accident, Melissa was unaware that many people around the world had learned about her through the internet, considered her "a medical pioneer," and were waiting with bated breath for the results. "I don't feel like a pioneer," she said. "I just hope I can recover, not only for myself but for others." Weighing her options after the accident, she had decided quickly: "OK, what have I got to lose?" Three months in an Israeli hospital ward populated mainly by youngsters paralyzed in traffic accidents was not how she had envisioned her first trip abroad. But the long stay in Israel was needed to give doctors a chance to evaluate her response and check for potential side effects, after which she could return home. The doctors told her it might take up to a year to know if the treatment had worked. By the time this book went to press in 2001, it was still too soon to tell whether Melissa would walk again, but no longer too soon to begin thinking that a treatment for paralysis would eventually be found.

Listening to Melissa talk about her decision to undergo a procedure never before tried in humans, I felt I was witnessing science history in the making. Over the years, many spinally injured people had traveled to distant lands for a chance of a cure, but Melissa's journey was different: The therapy she received belongs to a new wave of regeneration treatments, tested in humans after being officially approved by such bodies as the U.S. Food and Drug Administration. Melissa's journey was all the more dramatic, considering that only a few decades ago, the biggest challenge in dealing with paralysis was not cure but survival.

Until recently, reversing paralysis after spinal cord injuries was not a major issue because injured people usually died fairly quickly. One of modern history's most famous death scenes occurred in 1805 when British admiral Lord Nelson was spinally injured by a bullet fired from a French vessel during the battle of Trafalgar. "All power of motion and feeling beneath my chest are gone," Nelson told his ship surgeon. "My Lord, unhappily for our country, nothing can be done for you," the surgeon replied. Nelson survived his injury for barely three hours; during World War I, 80 percent of spinally injured soldiers died within two weeks. A 1936 surgery textbook stated: "In times of war, injured with a complete severance of the spinal cord should be enabled, by a timely transfer, to see their relatives before their certain death." People who survived rarely lived longer than two to three years and often died neglected, in the corners of hospital wards. Their two main scourges: urinary infections and pressure sores that developed on the skin from lying down or sitting in the same position. Ludwig Guttmann, a Jewish refugee neurosurgeon from Nazi Germany, described such patients as "the human scrapheap." It was thanks to Guttmann and a few other determined physicians—as well as the advent of antibiotics, which did away with many formerly fatal infections—that the fate of people with spinal cord injuries was to change dramatically toward the end of World War II.

Guttmann, a domineering man of immense energy, headed the famous unit for spinally injured people created at the Stoke Mandeville Hospital near London in preparation for the opening of the second front. There, "the German doctor" instilled a Prussian-Army-like discipline among the staff to maintain the highest standards of care for the paralyzed soldiers. "No patient was left lying in one position for more than an hour, no catheterization was performed without fully sterile procedures—and we began to see people surviving," recalls an observer. Just then penicillin became available. Lady Florey, the wife of Sir Howard Walter Florey, one of penicillin's three discoverers, came to Stoke Mandeville from Oxford once a week with minute quantities of the precious drug, which was still extremely scarce. Guttmann, later knighted by the Queen, also helped introduce sports for the disabled, emphasizing their right to be active members of society. Asked to sum up his rehabilitation philosophy in one sentence, he said: "To transform a hopeless and helpless spinally paralyzed individual into a taxpayer."

Elsewhere, medical professionals were reversing the fate of paralyzed people during the same years. In the United States, among the most prominent figures were Donald Munro, a neurosurgeon in Boston, Massachusetts, who pioneered a new approach to bladder care and promoted professional reintegration of people with spinal cord injury, and Howard Rusk in New York, who after the war created the institute for rehabilitation medicine that now bears his name. In Toronto, neurosurgeon Harry Botterell and physician Albin Jousse set up a specialized unit for Canadian veterans returning from World War II with spinal cord injuries. In Montreal, a similar unit and later a rehabilitation school were created by Gustave Gingras, who also helped introduce programs for spinally injured people in different countries, describing himself as a "traveling salesman in rehabilitation."

Thanks to increasingly effective antibiotics, new diagnostic tools, support technology, and surgical techniques—surgery is often performed after injury to relieve pressure on the cord or to stabilize the vertebral column—survival after spinal cord injuries continued to improve steadily after the war. By 1980, more than 90 percent of spinally injured people were surviving. Their life expectancy is still, on average, 10 percent less than that of able-bodied people, but for the first time in history, they make up a large and growing population.

## Bringing Hope Home

Spinal cord regeneration and repair might never have gotten under way without the lobbying of spinally injured people themselves. Starting around the early 1970s, some people with paralysis became actively involved in promoting research; no longer satisfied with being taught how to live in a wheelchair, they demanded a cure. The National Paraplegia Foundation in the United States began supporting spinal cord regeneration studies, and several of the Foundation's members launched a separate fundraising enterprise for the first time focused on "cure" research. Government funding for spinal cord research was raised, particularly in the United States where many veterans were returning from Vietnam with injuries to the brain and spinal cord. Unexpectedly, one of the greatest sources of inspiration in this lobbying was the landing on the moon. Spinal cord injuries present a more daunting challenge than the moon mission: Sending humans to the earth's natural satellite—not to mention bringing them back—required solving enormous but well-defined tasks, whereas trying to repair the nervous system involved, and still involves, grappling with fundamental unknowns. But the moon landing became a metaphor for turning a far-out cause into a victory.

It is not without irony that during the 1970s a number of spinally injured Americans turned for hope to the country that was losing the space race to the United States. Scientists and doctors in the Soviet Union, unlike their colleagues in the West, were actively searching for a paraplegia cure and developing new rehabilitation methods, almost none of which made the crossing to the West. Kent Waldrep, a Texas Christian University athlete paralyzed during a college football game, traveled to Leningrad in 1978 to receive treatment, including effective physical therapy not yet practiced in the

West. When he returned, American doctors found that he had regained some use of his paralyzed body but wrongly predicted that the improvement would not last. Waldrep responded by creating one of the first private organizations for funding regeneration research. "I went to Russia for hope. It's time I bring some of that hope home," Waldrep said.

At about the same time, several private foundations were formed in different parts of the world to stimulate the search for a cure. These undertakings were originally minor compared to the larger, more established efforts that emphasized better care for the disabled, but gradually, the "cure" contingent gained ground. In 1978, the Spinal Cord Society, still a highly active group, was created in the United States with the motto "Cure—Not Care" and a logo showing an X across a wheelchair symbol. In 1980, the foundation now known as the International Spinal Research Trust was created in the United Kingdom. One of the largest private organizations, the American Paralysis Association, later renamed the Christopher Reeve Paralysis Foundation, was founded in 1982. In 1985, a spectacular, are-you-out-of-your-mind kind of initiative was undertaken by the Canadian Rick Hansen, paralyzed in a car accident 12 years earlier. In a two-year marathon, he circled the world in his wheelchair to raise awareness and funds for spinal cord research and rehabilitation, covering some 25,000 miles across 34 countries, and using the funds he raised to form the Rick Hansen Man in Motion Foundation.

The foundations created science advisory councils that reviewed applications and chose the studies worthy of being funded, making sure that their money went toward scientifically sound research. At the beginning of the twenty-first century, dozens of voluntary organizations support research in different countries, and of these, several have come together to found ICCP, the International Campaign for Cures of Spinal Cord Injury Paralysis, coordinating their efforts worldwide. In what has probably been a unique phenomenon in the history of modern science, these and other "patient" initiatives helped revive spinal cord regeneration as a credible research field.

People with paralysis are generally not particularly interested in supporting pure, curiosity-driven research; scientists, on the other hand, don't like being told what to do: Guidance from above, they tend to feel, stifles creativity and is ultimately thwarting. But because the two sides are truly in need of one another, this partnership has worked amazingly well. Philanthropic support supplements government funding: For example, private foundations may support young investigators who have difficulty obtaining money for research, or they fund long-shot, exploratory projects that do not yet qualify for government grants. In addition to raising private funds, the spinal cord lobby has helped increase government research spending in a variety of creative ways: most recently, for example, by initiating laws, passed in different countries, whereby a proportion of fines for traffic violations is channeled toward research into nerve trauma. And last but not least, spinally injured people have "brought hope home" in a more literal way: They have been instrumental in reversing the nay-saying attitude toward the study of spinal cord regeneration and repair.

## On the Verge of a Nervous Breakthrough?

*Regeneration* has a different meaning in the nervous system than in other tissues. When such organs as the skin, bone, or liver regenerate, their cells divide, at least partly repairing the organ. In contrast, adult nerve cells do not divide; when they regenerate, this means that they regrow their long extensions, or fibers. But mere regrowth is insufficient for effective repair. In the spinal cord, repair implies restoring intricate circuitry, which in the human cord involves many millions of nerve cells and fibers. Simply suturing together the two segments of the spinal cord, as surgeons do with skin, doesn't re-create nervous circuitry. Thus, after the cord is injured, some of the damaged fibers must not only regenerate but also reorganize themselves into new, functioning circuits capable of conveying information.

This is precisely what happens when damage is inflicted on the brains or spinal cords of "lower" species such as fish and some amphibians and reptiles, the creatures that seem to be equipped with built-in repair mechanisms that kick in whenever their central nervous systems get into serious trouble. Fish with severed spinal cords regain their ability to swim, while certain amphibians and lizards even regenerate an entire tail, containing fibers from the spinal cord, their injured nervous circuits regenerating and effectively compensating for the damage. Not so, however, in mammals, including ourselves. In humans, repair occurs only in peripheral nerves, so called because they lie *away* from the brain and spinal cord, connecting the central nervous system to the rest of the body and controlling our muscles, glands, and internal organs. Cut a nerve in the arm, leg, or buttock, and it will usually grow back to fix the injury at least partly; that is why surgeons can often reattach severed fingers or toes—or even a penis—while restoring some sensation and function to the organ. But when the human spinal cord is injured, damaged nerves do not grow back; the result is permanent paralysis.

No one knows why, though regeneration researchers are struggling to make evolutionary sense of the conundrum. According to one hypothesis, an injury to the brain or spinal cord leaves mammals too sick or vulnerable to survive. "There would never be sufficient time to get repair completed before you die, so why bother building a repair system?" says Canadian neuroscientist John Steeves. In line with the same hypothesis, American researcher Wise Young notes that animals with regeneration capabilities tend to be small in size, and their nerves need to regenerate about an inch, a task that can be accomplished in several weeks. "It is perhaps possible for a small animal with a gene for regeneration to last that long and pass on this gene to offspring," Young says. As a result, natural selection would favor the survival of small animals with the regeneration gene. In contrast, large mammals, like us for instance, must regenerate nerves over long distances and would probably die before getting a chance to reproduce. Another theory states that mammals are too smart and too dependent on learned behaviors to have regeneration; a brain injury would erase these learned behaviors, leaving the animal as helpless as a newborn after repair. "There is no developmental pressure for such a repair mechanism to exist—instead, the brain and spinal cord are encapsulated in very strong bone, and we're simply not designed to go around in cars and on motor bikes," says Swedish neuroscientist Lars Olson. He adds that there may be no room physically for a regeneration system in the developed mammalian brain. "If we were as smart as we are and were also capable of regeneration, our heads might be twice as big as they are, which wouldn't be compatible with birth."

Despite all the unknowns, and the scale of the challenge, even skeptics believe it *may* be possible one day to reverse paralysis—not bad for a field so recently considered hopeless. The British neuroscientist Geoffrey Raisman, whose views are representative of the more cautious group, describes the current feeling as akin to being "in the cabin next to the engine in a boat—there's a rumbling, you can hear the wheels of history turning. If we can reconstruct severed nerves in the brain and spinal cord, then the blind will see, the deaf will hear, and the paralyzed will walk. Are we on the edge of doing those things? We'll only know when this happens." More hopeful scientists say effective spinal cord repair is only a matter of time, a question of when, not if, although most are wary of making exact predictions. Says Lars Olson: "It seems that, for man, nothing is impossible in the very long run. Spinal cord regeneration is one of the most difficult tasks, and there may be no major clinical breakthrough for another 10 years or so, but I hope I'm wrong—it could also happen next year."

# I
# THE SCIENCE OF THE
# IMPOSSIBLE

# 1

# The Maestro and the Dogma

A bout a century ago in Madrid, a legendary neuroscientist known to the world as Cajal, aided by his faithful disciples, produced one of the most extensive studies ever done on nerve degeneration and regeneration. It was not a study performed with the goal of treating patients. Santiago Ramón y Cajal was fascinated by the nervous system and wanted to reveal its structure, which was still mysterious and controversial.

For neuroscientists around the world, Cajal, one of the founders of their discipline, is a figure of nearly mythic proportions. He was born in an isolated Spanish mountain village and as a child dreamed of becoming an artist, but his physician father, desperate to rein in the rebellious boy, enrolled him as an apprentice to a barber and then a cobbler before sending him to medical school. Cajal, who would move through a succession of provincial posts to a professorship in Madrid, was so prolific and formulated such forward-looking ideas that many neuroscientists today feel they are still developing his concepts. His drawings of the nervous system, which rival the images obtained with today's most powerful microscopes, are still used to illustrate scientific talks.

The unusual circumstances of Cajal's rise to international fame have contributed to his mystique. In his time, most great scientists came from the educated elite and worked in countries where economic conditions favored the flourishing of science, which was far from true in Spain. In the early 1900s, Spain was going through an identity crisis after having lost Latin America and most other parts of its overseas empire in the preceding century; morale was low and the economic situation catastrophic. A movement of intellectuals known as "the generation of 1898," named for the year of the Spanish-American War, in which Spain gave up most of its leftover colonies, was calling for a national revival that would replace the country's lost imperial glory with intellectual and cultural prominence. Cajal, a village boy who grew up to become Spain's first Nobel laureate, was a perfect symbol for such a revival. He turned into a national hero; portraits of the short, wide-shouldered scientist, with an olive complexion and large-featured face, adorned stamps and money bills; streets were named after him and statues of him were placed in city parks,

As Cajal's ideas refused to become outdated over the years, admiration for his genius reached such proportions that some neuroscientists now feel the Cajal "cult" must be kept in check. Hero worship in science, these researchers feel, eclipses contributions made by less prominent scientists. In the case of regeneration, however, Cajal's fame came at a price: Probably due in part to his cult-hero status in neuroscience, he became known, rather unfairly, as the father of the dogma that effective regeneration cannot take place in the brain and spinal cord. In fact, like many of his contemporaries in other countries, what Cajal did was to document the *lack* of such regeneration. He never took a consistent stand against the *possibility* of central nervous regeneration, and his laboratory, amid its vast body of work on the topic, actually conducted experiments that would later serve as a launching pad for the current revival of spinal cord regeneration research. Still, it is the Spanish maestro who today commonly gets the blame for having shrouded the regeneration field in unwarranted pessimism for many decades.

## Revolution in an Italian Kitchen

Cajal embarked on regeneration research when his life's work was in danger. What mattered to him most throughout his career was to prove that the nervous system consisted of individual cells. Today it is hard to imagine that this basic notion could have been questioned, but in Cajal's time it was a radical new theory, attacked by scientists who believed that nerves in the body formed a continuous network.

By the middle of the nineteenth century, all living tissues had been generally accepted to consist of tiny compartments known as cells. The cells were clearly visible under a microscope after being dyed or, in technical language, "stained," with special chemicals. However, the structure of the brain, spinal cord, and the rest of the nervous system remained a mystery because none of the stains worked well for nerve tissue. Nerves appeared to early microscopists as an unwieldy mess of seemingly unrelated cell bodies and fibers.

The nerves literally came out of obscurity in 1873. That year, a 30-year-old Italian physician, Camillo Golgi, working in a small kitchen converted into a laboratory, discovered a method that revolutionized the study of the nervous system. He found that if a piece of the brain hardened in a chromate fixer is sliced up and dipped into a weak solution of silver nitrate, some nerve cells are stained brownish black in their entirety. (Golgi's discovery of the *reazione nera*, or "black reaction," as well as the subsequent advent of improved investigation methods, proved so central to the continuous progress in brain research that one neuroscientist, paraphrasing the lyrics from *My Fair Lady,* would later quip that "the gains in brain are mainly in the stain.")

Golgi was one of the major proponents of the idea that extensions of nerve cells in the body fuse into a network, much like pipes of a water system. The nerve fibers, he believed, "gradually lose their individuality while dividing to become extremely fine filaments." Ironically, Golgi's own method provided crucial ammunition for the opposing camp, supporters of the neuron theory. In the late 1880s, Cajal improved the black reaction and used it to conduct extensive studies of the nervous system, which laid the anatomical foundations of the neuron theory. The theory, formulated in 1891 by the German anatomist Wilhelm Waldeyer, states that the neuron, or nerve cell, is the elementary unit of the nervous system. Nerve cells, Cajal wrote, "touch each other like the foliage in a wood" and communicate through points of contact, now called synapses. Cajal referred to these points as intercellular "kisses," which "constitute the final ecstasy of an epic love story."

How could Golgi and Cajal look at the same tissues and see different things? One saw a continuous network, while the other saw individual units in tight contact with one another. (Today it is known that the contact between nerve cells is made across a minuscule gap, roughly the width of a human hair split 20,000 times, which could not possibly have been observed in the time of Golgi and Cajal.) The disparity reflects a constant tension between fact and theory in scientific research. Supposedly objective facts often depend on such fickle factors as the quality of tools and the skills of the experimenter, and without an underlying theory, scientists often find it difficult to make sense of murky findings. (Albert Einstein, once told that only observable entities must be used in formulating a theory like that of relativity, reportedly responded: "Possibly I did use this kind of reasoning, but it is nonsense all the same. . . . It is the theory which decides what we can observe.") Golgi, who worked mainly with adult brains, may have held on to his views on the network or, as it was then called, reticular theory, despite overwhelming factual evidence against it, because he could not imagine how nerve cells could operate in isolation; the network, he said, was "a postulate necessary to explain nerve transmission." Cajal, unlike Golgi, chose to do much of his work on embryonic brains—a choice that went along with his commitment to the concept of neurons as separate entities: In the embryo, nerve fibers are not yet covered with a protective fatty sheath that interferes with the Golgi staining, and nerve cells, still relatively small, are easier to observe separately.

In contrast to all other cells in the body, nerve cells have such a meandering shape it is little wonder their structure was difficult to figure out. From the core, or cell body, of each nerve cell extend treelike branches called dendrites (from the Greek for "tree"). Neurons in various parts of the nervous system differ greatly in the size and shape of their dendrites. Each cell also has one extension that is usually longer than the others, called the axon (from the Greek for "axis") or simply a nerve fiber. Some axons, particularly those in the brain, are only a fraction of a millimeter long, while others, such as the ones connecting the spinal cord with the feet, reach 30 inches (about 1 meter) or more in length. The dendrites and axons of different cells intertwine, granting nerve tissue a tangled appearance that Cajal described as an "inextricable thicket." (What we refer to as a "nerve" is often a multitude of axons bundled together like a cable. The human optic nerve, for example, which transmits visual information from the eye to the brain, is about 4 millimeters thick and contains about 1 million axons, whose cell bodies lie in the retina.) A neuron's job is to process information and convey it to other cells. In 1892, Cajal, in parallel with the Belgian anatomist Arthur Van Gehuchten, revealed how neurons are put together into circuits along which information travels: An impulse is picked up by the cell body or the dendrites and passed on further by the axon, so that the nerve cell works as an input-output unit and always conducts its messages in one direction. Synapses are thus established between cell bodies and axons, or between dendrites and axons.

## "Triumph of the Right Cause"

Just as it seemed that the neuron theory Cajal favored was prevailing over the network theory of which Golgi was a champion, controversy flared up again, peaking between 1900 and 1904, when scientists in the network camp proposed misleadingly attractive ideas about nerve growth and regeneration. Cajal would later describe the situation in military terms; he referred to the network theory as "a formidable enemy" and "an epidemic that was spreading and threatened to infect all minds" and called his regeneration research "a campaign" and "a hard battle on behalf of the truth." "It became fashionable to execrate and even smile at the neuron concept and at the theory of connection by contact," he wrote. "In the face of the crushing tide of error and in view of the repeated requests of my friends, I found myself compelled to halt in my path and descend to the arena."

Cajal had been studying regeneration for about a year when, in 1906, his research was interrupted by a telegram from Stockholm announcing that he had been awarded the Nobel Prize in physiology or medicine. In one of the most bizarre pairings ever produced by the Nobel committee, he was to share the prize with his scientific arch enemy, Camillo Golgi. "What a cruel irony of fate to pair, like Siamese twins united by the shoulders, scientific adversaries of such contrasting character!" Cajal would write in his autobiography. The prize committee took no sides in the nerve cell debate and diplomatically honored both scientists "for their work on the structure of the nervous system."

The meeting of the two laureates in Sweden brought no reconciliation, either personal or scientific. Cajal went to welcome Golgi at the Stockholm train station, but according to one account, "the ice was not broken either that evening or in the subsequent unavoidable encounters during their stay in Scandinavia." Golgi had resented having his name linked to that of Cajal long before the two scientists were jointly awarded the Nobel Prize. According to one of his biographers, Golgi, an excessively reserved and modest man who had worked in obscurity for many years before receiving recognition, apparently did not appreciate Cajal's meteoric rise to fame, which had been speeded in part by the Spanish scientist's direct and aggressive style.

The Nobel ceremonies culminated with Golgi committing one of the most famous faux pas in science history: He chose his award lecture as an opportunity to publicly demolish Cajal's views. He opened the lecture by declaring that the neuron doctrine was "generally recognized to be going out of favor" and went on to denounce the doctrine while, according to Cajal, "all the Swedish neurologists and histologists looked at the speaker with stupefaction." Golgi probably underestimated the popularity of the neuron theory. By the early twentieth century he had become a senator and reduced his research activities. It is also possible, according to his biographer, that he "feared an attack from Cajal and wanted to deal a preemptive strike." The following day, Cajal delivered his award lecture, in which he laid out the evidence in favor of the neuron theory. (Presentation of the 1906 Nobel Prize has been theatrically re-created twice at scientific conferences, in 1985 and 1999, with two neuroscientist members of the Cajal Club—an informal group of neuroscience history aficionados—playing Golgi and Cajal in period costumes and delivering excerpts from their Nobel speeches.)

Regeneration touched the heart of the dispute over the structure of nerves because the network and the neuron theories dictated two radically different regeneration scenarios. Supporters of the network camp argued that after injury, degeneration spread through the nerves and that the sheath of the damaged nerve fiber manufactured chunks of a new, healthy axon. These chunks supposedly reassembled themselves and fused into a continuous fiber (much like pieces of glass coming back together when a videotape of a smashed vase is played backwards).

Cajal was "thoroughly revolted" by this explanation. It implied that at different stages of its life, the organism "uses two different and almost antagonistic mechanisms for the construction of the nerves." In other words, during the development of the embryo, dendrites and the axon grow out from the cell body, but during regeneration the same axon is supposedly formed by fusion of independent segments. This did not make sense to Cajal. "Nature always proceeds in its operations with a spirit of strict economy," he said. The neuron theory dictated the following course of events: When an axon is cut, the segment that is separated from the cell body, the source of the axon's vital nutrition, quickly degenerates and dies, leaving behind an empty sheath. The surviving segment, the one attached to the cell body, then begins to sprout, like a young twig growing in place of a cut-off tree branch, and eventually restores the nerve fiber to its original length.

Cajal launched his regeneration studies when he was in his early 50s, and he referred to these years as "the zenith of my scientific career." For one thing, he was no longer a lonely genius peering into a primitive microscope. The Spanish government had created for him a special facility, called the Laboratory of Biological Research, in a wing of a gray stone mansion with a facade modeled after a Greek temple at the far end of the Atocha, a busy commercial street leading to the heart of old Madrid. Within a few minutes' walk up the Atocha was the University of Madrid's San Carlos Faculty of Medicine, where Cajal taught and also conducted research. The Faculty, today the quarters of Madrid's medical association, is a massive neoclassicist edifice built around a magnificent amphitheater where students watched corpse dissections. Cajal used both facilities, which were fitted with the best equipment available at the time, as a base for building up an entourage of students and disciples.

Around the time Cajal received the Nobel Prize, he got so busy that he entrusted much of the regeneration research to Francisco Tello, his best student and first true disciple. During the first two years of their regeneration research, Cajal and Tello revealed facts that Tello described as "most embarrassing" for the network theory. They demonstrated that the nerves regenerated precisely in accordance with the neuron theory scenario. Using a new technique developed by Cajal for seeing nerve cells in fine detail, they also discovered striking proof that regeneration recapitulated the embryonic growth of nerves. The regenerating sprout ended in a conical tip resembling a live creature, "a living battering ram" with hairlike projections that seemed to feel out the way for the nerve. This tip was the growth cone, whose initial observation in embryonic nerves had been one of Cajal's most important discoveries.

Cajal described his victory as the "triumph of the right cause," yet opposition to neuronism persisted. Neuroscience historian Edward Jones notes that in one case, this opposition involved a peculiar paradox. One of Cajal's main competitors in regeneration studies was Golgi's nephew, adopted son, and favorite pupil, Aldo Perroncito. Working in Golgi's laboratory in Pavia, in northern Italy, Perroncito made many of the same observations as Cajal and in several cases was ahead of the Spanish scientist, much to Cajal's concern. In studies published between 1905 and 1909 Perroncito showed, for example, how in a cut peripheral nerve, the stump attached to the cell body survived and sprouted new branches, while the other stump, the one severed from the cell body, degenerated. Yet surprisingly, says Professor Jones, this research conducted in Golgi's own laboratory did not seem to influence the Italian scientist's thinking on the neuron theory.

Debates over the neuron theory continued to rage in other areas of nerve study throughout the first half of the twentieth century. Only in the 1950s, when electron microscopes, which have a resolution thousands of times greater than the light microscopes, were applied to the study of nerves, was the network theory finally put to rest, although one last pocket of its diehard supporters, which persisted, of all places, in Spain, was eliminated as late as in the 1960s.

## A Historic First

Cajal was obviously fascinated by regeneration beyond the need to prove the neuron theory and devoted "six years of persistent investigations" to the problem. He described his own work and that of his students, as well as related studies of his contemporaries and competitors, in a two-volume treatise, *Degeneration and Regeneration of the Nervous System,* published in Spanish in 1913 and 1914 with money provided by patriotic Spanish physicians in Argentina, who wished to honor Cajal for his receipt of the Nobel Prize. The book, translated into English in 1928, is now seen as the first major attempt to define the issues involved in nerve regeneration in modern scientific terms.

The first volume of Cajal's treatise describes the successful regeneration of peripheral nerves. The second volume deals mainly with the central nervous system and its failure to regenerate effectively, and this text has been the source of glum quotes that won Cajal the dubious honor of fathering the dogma about the lack of regeneration in the brain and spinal cord. Even though the dogma had existed before, it became cloaked with authority in the wake of his work on the topic. Cajal wrote that his and other scientists' studies had confirmed that "as is well known, the central tracts are incapable of regeneration." He followed this by an often-cited description: "Once development was ended, the founts of growth and regeneration of the axons and dendrites dried up irrevocably. In adult centers the nerve paths are something fixed, ended, immutable. Everything may die, nothing may be regenerated."

Cajal, however, introduced the concept of "abortive" regeneration in the central nervous system. After injury, interrupted fibers in the brain, and particularly in the spinal cord, produce new sprouts with growth cones, which wither after a few days. Cajal asked a question with which regeneration researchers are grappling to this day: "Why, once the reconstructive movement is initiated, do the nerve sprouts lose their energy and suspend their growth?"

Cajal's major theory, "the banner under which we are working," was the idea, also most relevant today, that in order to grow, the new sprouts need stimulating substances that can feed them and guide them to their targets. He suspected that surrounding tissues release chemicals that attract axons and direct their growth. Cajal developed this theory by watching how the hairy, conelike tips of axons, the growth cones that he discovered, grope around, overcome obstacles, and beat their way to their new homes as if impelled by an invisible force. The nourishing and guiding chemicals would remain "invisible" for years, but when both types of substances came to be discovered in the second half of the twentieth century, they would give rise to two huge research areas in modern neuroscience, both pertinent to nerve regeneration. (The nourishing chemicals are now known as growth factors, while the area of research dealing with the guiding chemicals is referred to as axonal guidance.) Amazingly, Cajal predicted the existence of these substances while relying on intuition. He supported his insights by designing experiments that demonstrated the guiding power of the invisible chemicals during regeneration of peripheral axons.

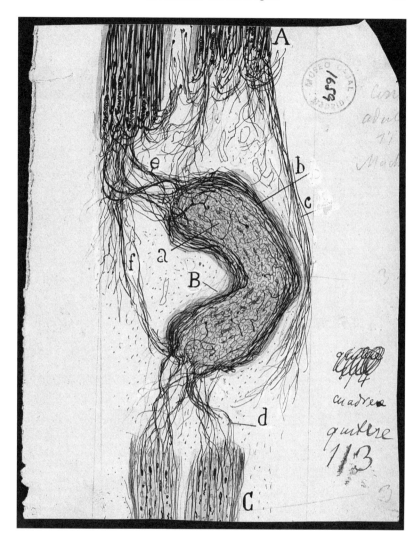

Cajal's drawing of a nerve regeneration experiment. Regenerating nerve fibers (e, f) from the injured sciatic nerve of a rabbit (A) are "attracted" by a transplanted piece of nerve (B) and follow an unusual, curved path toward the transplant. (Instituto Cajal, CSIC, Madrid, Spain.)

In these experiments, Cajal and his coworkers placed the target tissue, an implanted piece of nerve, in an unusual position, for example, not directly in the path of regenerating axons but on the side. The growing fibers changed their course, sometimes even making a U-turn, and implacably headed toward the implant. Cajal believed that in peripheral nerves, the growth-stimulating substances were secreted by the cells called Schwann cells. He hypothesized that in the brain and spinal cord, regeneration failed to occur because of the absence of Schwann cells. The lack of regeneration in the central nervous system, he said, was not "an irrevocable and unavoidable law, but a secondary consequence of the unfavorable chemical milieu." Much of the work on this theory and other aspects of central nervous regeneration was conducted by Francisco Tello, Cajal's right-hand man and anointed heir.

Tello came from the same part of northern Spain as Cajal, the austere mountainous region of Aragón. He had started to study surgery and was looking for an internship that would leave him enough time for his course work. He joined Cajal's laboratory temporarily because it had an opening that nobody wanted, but his supervisor's enthusiasm was so contagious that he soon decided to devote himself to basic science. Tello was a reserved, self-effacing man, who loved music and theater and was an enthusiastic photographer. His most striking trait was impeccable integrity. In teaching, his sense of justice translated into extreme strictness with students, who nicknamed him the "madman of San Carlos"; once he even failed a student who had come recommended by King Alfonso XIII himself. Cajal had serious squabbles with some of his other disciples, but his relationship with Tello, which lasted more than 30 years, was one of mutual commitment and trust. In their later years, during Tello's admission ceremony to Spain's National Academy of Medicine, Cajal would recall being impressed that such a brilliant student had opted for basic research rather than seeking what was then a much more prestigious career as a clinical professor or surgeon.

Tello seems to have harbored nothing but loyalty and devotion to the maestro, with never any resentment about Cajal's dominating presence in his life, but today many scientists feel Tello never received the recognition he deserved. "Unquestionably, the extraordinary fame of the maestro projected both light and shadow on other Spanish savants," said a speaker at a 1980 meeting dedicated to the centenary of Tello's birth. In those days, an assistant professor often got a chance to move up after the senior professor retired, but later in life, Tello's career was interrupted by politics. After Cajal's death in 1934, Tello was appointed director of the Cajal Institute, created on the basis of the Laboratory of Biological Research. However, being an agnostic and a republican, he was ousted from this post in 1939, when church-abiding conservatives came to power after the Spanish civil war, and was never able to resume productive scientific work. Tello made classic contributions to the study of the nervous system, but because he mostly worked in Cajal's laboratory, much of his research, including his landmark contributions to regeneration, tends to be obscured by his teacher's gigantic reputation.

In the second volume of his *Degeneration and Regeneration*, Cajal describes Tello's experiments, originally reported in 1911, that 70 years later would initiate the current revival of spinal regeneration research. Tello transplanted pieces of the rabbit sciatic nerve, a peripheral nerve from the leg, into different parts of the rabbit brain—the cerebral cortex, the cerebellum, and the optic nerve. Like a vine that climbs up a wall or creeps along the ground, a growing axon needs a surface in order to grow. A piece of peripheral nerve was supposed to provide such a surface for the cut axon, but also a "chemical milieu" favorable for growth, rich in stimulating substances that were missing in the central nervous system. Tello would later say in a memorial lecture that launching these experiments was his idea but he acknowledged that the transplants were neither his nor Cajal's invention. He noted that several of his contemporaries had produced "a change in regenerative acts" by transplanting embryonic tissue, Schwann cell sheaths, or other "tissues capable of stimulating growth" into the brain and spinal cord. No one, however, had seen signs of convincing axonal regeneration.

In Tello's experiments, cut brain fibers did grow back, and this regrowth did not abort after a few days as when brain cells tried to regenerate alone. Tello used staining methods that revealed individual axons and saw that the brain axons turned toward the transplant and grew through the grafted 1-centimeter pieces of peripheral nerve. Unlike studies by all earlier researchers and physicians who, throughout many centuries, had tried to treat the brain and spinal cord but measured such vague parameters as general improvement, this was the first scientifically documented evidence of the possibility of central nervous regeneration. It was not an irrevocable proof—the technology for furnishing such a proof was still lacking—but it sufficed to plant the seeds of the future search for successful regeneration. Cajal described Tello's results with his usual flamboyance. In the presence of the transplants, he wrote, central nervous fibers "shake off their sluggishness, become turgid with activity and give off very long sprouts, which assail the nerve implant with the same aggressiveness and strength of growth that is characteristic of the sprouting of the interrupted sciatic nerves." Cajal concluded that his own and Tello's findings "definitely refute the generally accepted dogma of the *essential irregenerability of the central pathways*."

As it happened with many ground-breaking findings from Cajal's laboratory, the results of these experiments—at least in the way they were interpreted by Tello and Cajal—were not believed until years later, when they could be confirmed with modern tools. The number of experimental animals was small, and—future researchers would reason—the growing sprouts could have crept in from peripheral nerves in the scalp rather than extending from the central nervous fibers in the brain. Tello's experiments were intriguing enough for scientists to repeat them in subsequent decades with varying degrees of success, but generally, the conclusions about the central nervous fibers' regeneration capacity were not taken seriously. In contrast, readily believed and enshrined in all twentieth century textbooks were Cajal's statements about the lack of regeneration in the central nervous system. These statements fitted in with what neurologists saw in their daily practice—injuries to the brain and spinal cord produced severe damage that medicine could not reverse. Shortly after the publication of Cajal's regeneration opus came World War I, whose numerous casualties reinforced the physicians' sense of helplessness in the face of central nervous system injuries.

It is hard to know whether Cajal himself, who was mostly interested in basic science, not clinical medicine, believed in the possibility of effective brain and spinal cord regeneration in humans. The contradictory statements in his book—for and against central nervous regeneration—may have reflected the mood swings from which he suffered in his later years and which may have swayed his opinion in opposing directions during the long process of working on his regeneration treatise. He certainly promised no cures. "We are still in the phase of collection of materials," he wrote while meticulously recording both negative and positive findings. Subsequent generations of scientists read Cajal's two-volume opus selectively and for many years ignored the positive evidence. Only decades later would they pick up on Cajal's forward-looking ideas, including a suggestion with which he concludes a long passage on the lack of central nervous regeneration: "It is for the science of the future to change, if possible, this harsh decree."

# 2

# Regeneration's Solitary Warrior

For more than 40 years after Francisco Tello's experiments, only sporadic attempts were made to regenerate the mammalian brain or spinal cord. A handful of researchers in the 1920s, 1930s, and 1940s tackled the topic: Some repeated Tello's transplants, others tried different approaches, but most obtained negative or ambiguous results; the rare positive findings were generally dismissed as a mistake or an illusion. The prevalent attitude toward regeneration during these decades reminded one researcher of a humorous aphorism: "A pessimist is one who views the world situation and is absolutely sure we're going to hell; the optimist isn't absolutely sure."

Making a claim to successful central nervous regeneration under these circumstances would have turned any scientist into a lone soldier taking on an entire army. When the American researcher

William Windle came out with such a claim in the 1950s, he didn't mind the challenge—in fact, he loved it. Dr. Windle would carry the banner of regeneration research almost single-handedly for more than 20 years. His personal integrity was never questioned, but most scientists did not believe he could regrow central nervous fibers. Windle fiercely defended his evidence, insisted that the field was worthy of serious study, and fought for money to do the work. "He was a bit of a rebel, and the fact that he would promote research that went against all established concepts was part of his personality—in other words, he rebelled against the entire scientific community," says his former student and later collaborator and friend Dr. Carmine Clemente.

The Indiana-born Dr. Windle may have appeared to be an unprepossessing Midwesterner but is, in fact, remembered by his peers as a scrapper and "one hell of a character." He was a dominating, no-nonsense type of man with a track record of being an effective leader. He also had a record of getting into trouble with the administration in every institution where he worked. Yet people who worked for him generally adored him. "He'd take on the universe for you, he really would," recalls a scientist who worked in his laboratory. Nobody ever called him Bill, only "Dr. Windle." He always wore a lab coat and a bow tie and spoke in a categorical tone that would have given him the air of an army general were it not for a mischievous twinkle in his eyes. The intensity of his emotional involvement in a subject could be measured by the redness of his face, as Windle, a rather small man with thin blond hair and light-blue eyes, would turn red up to his balding head whenever he was excited or upset. Faced with his crimson anger, people who did not know him well would quickly retreat, worried that he was about to have a fit. Once they were out the door, he'd break into his hearty laughter, adding, "We got them, didn't we?"

Windle's knack for generating controversy didn't stop him from becoming a world authority in his field, the development of the central nervous system in the fetus. He became a professor at Northwestern University's Medical School on the university's Chicago campus, where he had done his graduate studies. Then, in

1947, he accepted the flattering offer to build up the anatomy department at the Ivy League University of Pennsylvania in Philadelphia, the oldest school of medicine in the United States. It was at Penn, when Windle was past 50, that an accidental observation in his lab changed his scientific life—and the future of spinal regeneration research.

In 1949, Windle and his team were engaged in a project that was not supposed to produce anything dramatic; it was aimed at establishing what part of the central nervous system was responsible for the control of fever. After creating injuries in different parts of the brain and spinal cord, the scientists examined the response of experimental animals to a fever-producing drug called Piromen, a complex sugar derived from *Pseudomonas* bacteria. Several animals whose spinal cords had been severed needed to have their bladders emptied manually because their hind limb areas were paralyzed. Working on this project was a young scientist, William Chambers, who had followed Windle from Chicago to Philadelphia. One day Chambers noticed that a paralyzed dog howled when its bladder was compressed, yet the paralyzed animal was not supposed to have any feeling past the injury site. Had its spinal cord healed sufficiently for pain impulses to be transmitted from the bladder to the brain? Chambers examined the dog's spinal cord under the microscope and, to his amazement, saw a considerable number of nerve fibers growing across the injury. The same was true for other animals treated with the fever-producing drug.

The discovery was so unusual that Windle, who was an excitable man, immediately translated the excitement into action. Together with Chambers and his new graduate student, Carmine Clemente, as well as other people in his lab, he launched an intensive research program to explore the new finding. Soon it looked as if what they had found was nothing short of a clue to solving the problem of spinal cord regeneration in mammals.

## Use of Animals in Research

Since I will be telling the reader about numerous discoveries based on animal experiments, this is perhaps an appropriate place for an aside on animal experimentation, a question that was on my mind throughout the work on this book. Opponents say experimenting on animals in the name of science is morally unacceptable. Meanwhile, all major medical advances that are now taken for granted and save millions of lives have at least partly resulted from animal research, including our ability to prevent or treat polio, diabetes, measles, smallpox, and heart disease. Insulin treatment for diabetics, for example, stemmed largely from research on dogs. Unfortunately, paralysis is no exception; to find a cure, scientists have no choice but to experiment on animals that are at least partially paralyzed.

Today animal experiments in many countries are strictly regulated, and in regeneration research, they are almost exclusively conducted on rats and mice. The small rodents are cheaper, more genetically uniform, and for reasons that are not entirely logical, scientists find it emotionally easier to work with them than with species that form a bond with humans. Permission to experiment on such species as cats, dogs, and monkeys is granted only when there is no other way to obtain the data needed to treat humans. All experiments must be approved by animal ethics committees, which require that the animals be treated in a humane manner, that they have company, and that as few animals as possible are used in each experiment. In laboratories where spinal cord research is done, paralyzed animals, like humans with paralysis, unable to empty their bladders voluntarily, receive help with bladder control. Care is usually provided by trained technicians or veterinarians, and the animals' health and well-being are closely monitored. And beyond other considerations, the animals must be treated well for the simple reason that experiments cannot be properly conducted with sick or overstressed subjects. In fact, the difficulty of providing paralyzed rats with proper care deters many scientists from embarking on paraplegia research.

In Windle's time, there were no animal ethics committees, just as there were no regulating bodies to strictly oversee experiments on humans. Windle was ahead of his time in the humane treatment of his laboratory animals; scientists who worked with him say anesthesia was always used, and the animals were treated with the same care accorded human patients.

## Milestone or Miracle?

The Piromen experiments caused Windle to believe that the explanation for what he called "the enigma of paralysis" was a physical roadblock—the scar. When tissues are injured, the body builds a scar that probably functions like a moat around the injury site, containing the spread of the damage. Windle and his colleagues found that the injured spinal cords of experimental animals also contained dense, fibrous clumps at the site of the injury that appeared to block any fibers trying to regenerate. In sharp contrast, the spinal cords of animals treated with Piromen had a loose, wispy scar that allowed the regenerating nerve fibers to pass through. In some unknown way, even when given in small doses that did not produce fever, Piromen appeared to create pathways in the scar tissue.

The report on the study, first released in 1950, was received by scientists with what Windle described as "skeptical interest." At one conference, a Boston neurologist requested that Windle and his collaborators add the words "in animals" to the title of their report on regeneration because, he said, "we all know that it never occurs in man." However, following a brief article by the Associated Press, Windle was flooded with letters from paralyzed people all over the country. He patiently answered them all that the treatment required further testing and that it was likely to help only newly injured people whose spinal cords had not yet formed impenetrable scars.

One letter, however, stood out as particularly heartbreaking. It was from the mother of a 19-year-old University of Georgia student, Charles VanDiviere, who had just broken his neck in a car accident while returning to college from a spring vacation. Doctors said that if he survived, he would be paralyzed from the neck down. "His injury was so recent that if there was a chance for anyone [to be helped by Piromen], he would be that one," Windle told a newspaper reporter. The student's doctor pleaded with Windle to give Piromen a try.

Those were the days before regulating agencies imposed strict control over experimental medications, and often the family's consent was sufficient to authorize the use of an unproven substance. Windle decided to take a chance, and 43 days after his accident VanDiviere received injections of Piromen. What happened next was, as one newspaper headline put it, either a milestone or a miracle. The handsome, 6-foot 2-inch young man began to recover and within two months regained bladder control and hand motion. In the following months he got back on his feet, returned to college, and resumed swimming and dancing.

Windle, a principled and scrupulously honest scientist, never claimed that Piromen cured VanDiviere. The youth's cord was badly crushed but not cut across, and Windle stressed that people with such injuries had been known to recover in the past. He added that the youth's parents could afford the best possible care. But when he brought VanDiviere before scientists of the National Institutes of Health in Bethesda, Maryland, the media ran wild with the story. *Time, Newsweek,* and the *New York Times*, among many others, carried articles about Piromen, and the tabloids said that thanks to the "wonder drug" textbooks might have to be rewritten.

On the crest of the enthusiasm in 1951, Windle moved to Morton Grove, Illinois, to become scientific director at Baxter Laboratories, manufacturers of Piromen, as the company was willing to invest in exploring the drug's new potential. However, when a dozen other spinally injured people received Piromen, as did several children with polio, the drug did not seem to be particularly helpful in humans. VanDiviere's recovery seemed to be a miracle after all. Large-scale human trials were postponed until animal experiments would produce a clearer picture.

## A Question of Making It Work

Windle's Piromen discovery coincided with a sweeping change in the fate of paralyzed people, who were surviving their injuries in growing numbers. The Korean conflict, which ended in 1953, left some 1,000 American veterans paralyzed, bringing new attention to the problem. These veterans joined about 2,500 Americans paralyzed in World War II. In addition, the growing use of private cars, whose sales soared after the war, resulted in thousands of civilians being spinally injured in car accidents.

The U.S. medical authorities decided to stimulate basic research that could lead to a treatment for paraplegia and other neurological ills, and Windle was offered the opportunity to create a Laboratory of Neuroanatomical Sciences at the government's National Institutes of Health (NIH). In February 1954, he left Baxter, and together with his assistant and their experimental animals, rode a double-length railroad car, converted into a mobile laboratory, from Morton Grove to Bethesda. While at the NIH, he would head research efforts in two fields: the effects of oxygen starvation on the brain of the newborn and regeneration in the brain and spinal cord. In the regeneration research, he continued animal experiments with Piromen and tried other approaches. These included injecting hormones into the injured spinal cord or, in collaborative studies with a New York neurosurgeon, inserting a Millipore filter at the injury site. Millipore, a porous plastic material originally designed to filter bacteria and viruses and also used to filter satellite and rocket fuel, was intended to ward off scar tissue and guide regenerating fibers in the spinal cord.

In May 1954, upon the initiative of his NIH boss, Pearce Bailey, Windle organized the first conference in history on regeneration in the central nervous system. The meeting brought together most of the scientists who had ever worked in this area, a total of 33, the majority of them from the United States. Today this number of researchers could fit into one large lab, and in those days it was not a large number either, but among them were several scientists who would later make classic contributions to brain and spinal cord repair.

One of them was a future Nobel laureate, the Italian scientist Rita Levi-Montalcini, who reported on an obscure substance, later to be called the nerve growth factor, that stimulated the growth of peripheral nerves in a laboratory dish. Other speakers reported on studies dealing directly with the central nervous system or animal studies that seemed far more relevant to the meeting's topic, but Windle must have sensed that a new and unusual nerve growth phenomenon could be significant. It would take three more decades for future research to prove him right.

As the talks at the conference moved up the evolutionary ladder from fish to frogs to mammals, there was less and less certainty about the prospects for spinal cord regeneration. That is, until it was the turn of notable neuroscientist Ralph Gerard, Windle's personal friend from Chicago and one of the greatest enthusiasts the regeneration field ever had. Gerard had done relatively little work on regeneration, but a project he carried out in 1940, based on Tello's experiments, was one of the first controlled studies to produce positive results. He proclaimed that his own research, as well as that of others, had disproved the belief that "some negative law of nature forbids regeneration" and climaxed his talk with the statement: "The exciting thing, I repeat, is that, once one knows regeneration is inherently possible, it is just a question of making it work."

Windle, who spoke last, discussed the extraordinary case of a Philadelphia waitress, Clara Nicholas, who had been paralyzed in a shooting some 50 years earlier, in 1901, at age 26. Her surgeon had performed a procedure, appealing in its simplicity, that would occasionally be attempted by other physicians in subsequent decades without ever being definitively proved effective: He stitched the two severed stumps of Clara's spinal cord together. (Stitching alone is unlikely to be helpful; for recovery to occur, axons must regenerate and restore communication between the brain and spinal cord.) Several months later, Clara made a remarkable recovery: She could flex and extend her legs, rotate her thighs, and even stand up. She could also feel heat, cold, and pain below the level of her wound. Philadelphia doctors went out of their way to explain Clara's recovery by anything but regeneration. One neurologist at the University of Pennsylvania even suggested that Clara probably had two spinal cords and that only one of them must have been cut by the bullet.

At the Bethesda conference and at subsequent scientific and public meetings, Windle would tell Clara's story over and over, noting that when she died, the autopsy revealed a single spinal cord. There was no evidence of regeneration but, Windle pointed out, neither was there any proof that regeneration had not occurred in the early months after the shooting. In the hospital, Clara had a severe bacterial infection with high fever, and Windle hypothesized that her body might have manufactured a substance akin to Piromen, which might have facilitated regrowth of fibers. Above all, the saddest thing for Windle was that no one was stimulated by Clara's case to initiate research. To him, she was a symbol both of the possibility of regeneration and of the way this possibility had been repeatedly killed by negative attitudes. He wound up the meeting with a call to take regeneration research from animal experiments to humans.

Today the Bethesda conference is looked upon as a historic event, an "official thaw of the age-old iceberg of doom." The book of its proceedings, edited by Windle, inspired an entire generation of future researchers. But at the time, Windle saw the conference as a flop. "Few of those who attended left with an urge to use any of their precious time in research in this unpopular field. Within five years, only four or five of the conferees were actively engaged in central nervous regeneration," he wrote.

Beyond scientists' inertia, there were objective reasons for Windle's difficulty in stirring up enthusiasm for regeneration. In the past as well as now, ignorant well-wishers and outright quacks have often promised to "cure" people of paralysis. As a result, scientists received every new promise of a paraplegia remedy with great suspicion. The science of regeneration was also problematic. When the spinal cord is cut, its nerve fibers make a terrible mess. When Windle got the nerves to regrow, his colleagues questioned whether these fibers indeed belonged to the cord or whether they had regenerated from the peripheral nervous system. Those were still the "dark ages" of neuroscience, when the scientists' ability to visualize the details of nerve tissue was extremely limited, and with technology available at the time there was no way of telling the fibers apart.

Moreover, Piromen, which remained Windle's greatest source of inspiration throughout his bid for regeneration, produced variable results, enhancing the mistrust of the scientific community. The original batch of Piromen seemed to have worked best, but the subsequent, more purified batches made for commercial distribution never worked quite as well, and other scientists had trouble repeating Windle's results. In science, the ultimate "seal of approval" is having the study reproduced by an independent laboratory. In most cases the scientists' credibility is not in question, but the practice rules out the risk that the study authors made a mistake or let their judgment be clouded by wishful thinking. Moreover, by repeating the same steps and arriving at the same results, an independent team confirms the cause-and-effect relationship between the study's methods and its findings. The Piromen findings never received this kind of approval. Another disconcerting factor was that no one had any idea how Piromen worked. Scientists suggested that it might be triggering the body's immune or hormonal response in a way that somehow facilitated regeneration, but they were never able to figure out how. (Piromen's mechanism of action remains unknown, but in the late 1990s the immune response, which Piromen might have affected, would be harnessed in the regeneration treatment that American teenager Melissa Holley would receive in Israel.) Lastly, none of the approaches tried by Windle made paralyzed animals walk. In the most successful studies, regenerating fibers grew for several weeks or even months and were able to conduct nerve impulses, but the growth eventually stopped without allowing the animal to recover significant function.

By the time Windle retired from the NIH in 1963, he was a maverick insisting on studying an impossible problem. After his retirement the government dropped its regeneration program, while Baxter Laboratories had stopped funding Piromen research. For part of the 1960s it was virtually impossible to obtain new grants for research on spinal cord regeneration. This state of affairs hit an all-time low after November 1967, when a surgeon in Toronto announced to the press that he had cured individuals with paralysis by a surgical procedure. The reports were almost immediately proved false, and spinal cord regeneration was increasingly branded as a fiefdom of charlatans.

# A Turning Point

After "retiring" from the NIH, Dr. Windle worked for eight years at New York University's Institute of Rehabilitation Medicine, founded by the famous physician Dr. Howard Rusk. Windle pursued his research on brain damage at birth, which had won him the prestigious Lasker Award in 1968, and carried on with his spinal cord studies. It was during this period that he gained new allies for his regeneration crusade—the paralyzed people themselves. In 1969, Windle got a call from Alan Reich, then president of the National Paraplegia Foundation, who had been paralyzed in a diving accident. Reich suggested convening a scientific conference on regeneration in the brain and spinal cord. "You get the scientists, I'll get the money," he told Windle. "I've been waiting 15 years for this call," Windle replied.

The conference, called "Application of New Technology to the Enigma of Central Nervous Regeneration," was held in February 1970 at the opulent hotel The Breakers in Palm Beach, Florida. It was the first major regeneration meeting initiated and attended by members of the paralyzed community and funded largely by private sources. At the beginning of the conference, according to one observer, "the only one there with any idea [that] paralysis was a solvable problem was Windle." But then different talks began to fit together like pieces of a puzzle, offering new avenues for studying regeneration with new techniques. The scar no longer looked like the major culprit, but other promising lines of investigation emerged, particularly the need to answer basic questions on various topics: growth-stimulating substances, molecular processes in the neuron, and flexibility, or plasticity, of the central nervous system. All these would loom steadily larger in spinal regeneration studies as the years went by.

At the end of the meeting, the National Paraplegia Foundation sent a telegram to President Nixon, demanding funding for basic research toward a cure for paraplegia. The telegram stated: "As a result of this conference, the participants are now convinced that regeneration in the central nervous system, formerly considered hopeless, is amenable to solution through such basic research."

Reich's next idea for reviving regeneration research was based on the premise that scientists have the same emotional needs as other humans. A relatively modest sum of money, he reasoned, may do wonders in stimulating the field if given as an award to a scientist whose contributions might otherwise go unrecognized. That was how the William T. Wakeman Award, the first regeneration prize and still considered one of the most prestigious in the field, was created. William Wakeman became paraplegic after being shot by his wife in a family argument in their elegant Palm Beach home. He was determined to walk again or die, and several months after the injury he entrusted himself to a surgeon who promised to cure him of paraplegia. As a result of the clandestine five-hour operation, Wakeman died. His guilt-ridden widow became active in supporting regeneration research, helped organize the Palm Beach conference, and later, on Reich's initiative, created the Wakeman Award. In 1972, Windle and Roger Sperry, the future Nobel laureate, became its first recipients.

## The Russians Are Coming

Shortly after the Florida conference, sensational news arrived from the Soviet Union. Researchers there claimed to have developed a successful treatment for paralysis. In 1974, at the age of 76, Windle convinced the U.S. medical authorities to send him behind the Iron Curtain to find out if the treatment was for real.

Upon his arrival in the USSR, Windle must have had the surprise of his life: He learned that the Soviet studies had been prompted by his own Piromen research. "The Russians gave more credence to our work with Piromen than did our own compatriots in the U.S.A.," he noted bitterly. Back in the 1950s, a member of the USSR Academy of Sciences had learned about Windle's work and asked local microbiologists to prepare a drug similar to Piromen. The Soviet scientists used tissue-dissolving enzymes to soften the spinal cord scar, in what has been described as a "meat tenderizer" approach. Windle's book from the Bethesda conference had been translated into Russian, and Soviet scientists carried out intensive regeneration studies throughout the 1960s, unbeknownst to their English-speaking colleagues.

In the USSR, Windle heard reports of humans treated with the enzyme therapy, but he was unable to tell whether the people had been helped by the treatment. Each human injury is different, making it difficult to evaluate the outcome of therapy, especially since some people with paralysis regain partial use of their bodies without any treatment. Besides, for obvious reasons, it is impossible to remove the human spinal cord for a post-treatment evaluation. At about the same time, several Americans with spinal cord injuries traveled to the Soviet Union and returned in better shape—they were unable to walk again, but some regained better use of their arms and upper bodies—yet here again, recovery may have been unrelated to the enzyme treatment. Unlike their colleagues in the West, Soviet doctors practiced an active rehabilitation approach, and it was impossible to know whether the people had improved as a result of rehabilitation or regeneration.

In contrast, Windle was impressed by the studies on rats conducted in the Soviet republic of Armenia and went to great trouble to obtain visas for the Soviet scientists to travel to the United States. They presented their findings at a conference in Florida and stayed on for collaborative research. However, when Dr. Lloyd Guth of the University of Maryland School of Medicine, Windle's loyal disciple and an unflinching early advocate of regeneration, repeated the Soviet animal experiments, he discovered what was clearly an honest mistake. "The Russians," as everybody called them, were cutting the spinal cords of rats by pushing the scalpel downward, and a few nerve fibers would sometimes remain uncut. Rats need very few spinal fibers to regain control of their hind legs, and these few spared nerves allowed them to perform what is known as "spinal walking," leg movements that are largely controlled by neurons in the spinal cord with minimal or no input from the brain. In Guth's experiments, six of the eight paralyzed animals began to walk within one month even though they had received no enzyme treatment.

Years later this finding proved of monumental importance when spinally injured humans were also shown to recover thanks to a few surviving fibers, but at the time Dr. Guth saw only the negative side. Before administering the enzymes, he inserted a wire underneath the spinal cord of a rat, then lifted the wire through the incision to make sure all the fibers had been cut. In this new round of experiments, all rats remained paraplegic.

## The Ultimate Battle

Windle was unscathed by the Russian fiasco. Another attempt at regeneration had failed, but others could succeed and he demanded that they be given a chance. By that time he had left New York University and moved with his wife, Ella, to Granville, Ohio, the site of Denison University, where he and Ella had met during their undergraduate studies some 50 years earlier. While in his late 70s, Windle obtained a small grant to study the link between regeneration and female hormones. Controversial recovery stories—remember Clara?—appeared to be more common in women than men. What if the female hormones somehow facilitated regeneration? These Denison University studies, however, never produced conclusive results.

Despite all the setbacks, Windle's impact on the regeneration field proved long-lasting. After the 1970 meeting in Palm Beach, the Florida regeneration conferences continued at approximately two-year intervals for more than a decade and later served as a model for other such meetings in the United States and elsewhere. Another important part of Windle's legacy is the journal *Experimental Neurology*, which he founded in 1959 and which continues to serve as a major forum for regeneration research. And last but not least, he inspired other scientists to join the field. In particular, he was exceptionally generous and supportive in promoting the careers of younger scientists, some of whom would help revive the study of regeneration.

This revival was already under way in the late 1970s and early 1980s, but it is hard to tell whether Windle, or anyone else for that matter, could see this at the time. It would take another 20 years for regeneration to turn into a bustling research field with scientific teams around the world racing to find a cure for spinal cord injuries. The advent of new technologies making it possible to peer into the innermost fabric of the brain and spinal cord would contribute to the revival of research. Windle died in 1985 at the age of 86, probably without knowing that he had won the biggest battle of his life, that of gaining respect for regeneration. But according to people who knew him, not for a moment did he ever doubt that he would.

# 3

# The Gains in Brain
# Are Mainly in the Stain

B ack when William Windle had launched his regeneration crusade in the 1950s, the very notion of change in the adult central nervous system was considered heresy. Throughout the
first half of the twentieth century, the system had been viewed as
rigid and hardwired; scientists thought that for the orderly flow of
information, nerve circuits in the brain and spinal cord had to remain fixed. The idea that an adult nerve fiber could alter its shape
and form new connections seemed as wild as the suggestion that
circuits could rearrange themselves in a stereo or a computer would
seem today. But during the years that Windle was fighting for regeneration, a true paradigm shift occurred in neuroscience: Scientists started seeing the brain and spinal cord as anything but rigid.
It is now agreed that these organs are changeable and malleable—

attributes that create a basis for repair and recovery. This flexibility is referred to as the nervous system's "plasticity."

The first modern plasticity study on the central nervous system was conducted by two of Windle's disciples, and it was initially spurned. At the University of Pennsylvania, Drs. Chan-nao Liu and William Chambers (the same Chambers who had made the original Piromen observation) were inspired by the knowledge that many people recover after stroke and even after certain types of partial spinal cord injury. But what is the anatomical basis of this recovery? Liu and Chambers hypothesized that recovery occurs because after injury, healthy nerve fibers send out short shoots, or sprouts, that compensate for the fibers that had been damaged. That this phenomenon, called "sprouting," occurs in peripheral nerves was well known. What Liu and Chambers showed in experimental animals was that after spinal cord injury, healthy nerve fibers sprouted near the injury site *within the central nervous system*, creating a dense meshwork that filled up the space freed up by the cut nerves. The nervous system, it seems, abhors a vacuum: When a portion of the spinal cord loses its nerve circuits, spared nerve fibers in the vicinity sense the void and make up for the loss.

The scientific community received the Liu and Chambers sprouting report, which appeared in 1958, with great suspicion. Sprouting, which refers to tiny, sometimes microscopic offshoots emanating both from injured *and* healthy nerve fibers, is different from "true" regeneration, which refers to the long-distance restoration of a *cut* axon to its full length. Still, even minor growth was not thought to take place in the adult central nervous system. Liu and Chambers used a newly developed and still rather crude method for tracing nerve fibers—the shape of the sprouts could only be revealed after the neurons were killed, producing a particular pattern of decay—and they had difficulty proving their findings were not a fluke. Other scientists tended to trust accepted theories rather than the emerging and imperfect tracing technology.

Liu and Chambers were truly ahead of their time, and their work might have been forgotten were it not for the revolution that was occurring in technology used for the study of the nervous system. More than 10 years would pass after their 1958 study before the appearance of the next milestone plasticity report—and this time the finding would be in the brain.

## Not Supposed to Happen?

One evening in 1965, Geoffrey Raisman, then a young lecturer at the University of Oxford, was conducting an anatomy tutorial that both he and his two students found exceedingly boring. His desk was strewn with images of a rat brain he had taken earlier in the week with the help of an electron microscope, at the time a relatively new tool in the study of the nervous system. Trying to enliven the dull two-hour session, Raisman invited his students to examine one particularly clear image and speculate on what it revealed about the workings of the brain. The picture had drawn his attention not only because of its high quality, but because it showed something that was not supposed to happen. On the electron microscope snapshot of the rat brain, Raisman saw that an adult axon had formed two connections, instead of the usual one, with an adjacent nerve cell.

The chance evening observation fascinated Dr. Raisman and prompted him to conduct a systematic study of two adjacent nerve networks in rat brains that have well-defined patterns of connections. He found that when either of the networks was injured, the other sent out sprouts that took over the vacated space and formed new, extra connections. The study provided the first convincing evidence that the central nervous system was not entirely hardwired after all; it could give rise to new circuits after injury.

Raisman hesitated for three long years before sending his iconoclastic findings to a scientific journal in 1968. That year many European universities were shaken by massive student protests, which escalated into a rebellion against societal norms. The "city of spires," as Oxford is sometimes called because of its skyline of Gothic towers and steeples, was untouched by the upheavals; the conservative student body of the 800-year-old university stayed clear of the riots. Yet Raisman's work was an instance of rebellious neuroscience. If these findings were a mistake—or if they were *perceived* as a mistake—announcing them to the world might have meant professional suicide. Many researchers were skeptical about the research, but Raisman gratefully remembers the support of others who encouraged him to make his work known, including his Oxford mentor who advised him, "Anyone who isn't prepared to make a mistake won't make a discovery."

Dr. Raisman's report, published in *Brain Research* in 1969, was convincing thanks to electron microscopy, which made it possible to see things about which scientists had speculated for years. Electron microscopes work on the same principle as light microscopes but are vastly more powerful. A beam of electrons is deflected as it passes near the atoms in the observed specimen, and the deflection is used to form an image—in the same way that glass lenses form images with deflected light. But because the wavelength of an electron is much smaller than that of light, the resolution of an electron microscope is thousands of times greater. It can reveal the minutest details of an object while enlarging it 1 million times—a magnifying power that makes a hair appear several feet wide.

Because of their fantastic resolution, electron microscopes settled many long-standing scientific disputes, including the last debates over the neuron theory: that the nervous system is made of individual cells. When observed through an electron microscope, the tiniest structures in nerve tissue, including the points of connection between nerve cells, the synapses, Cajal's intercellular "kisses," are clearly visible. In his 1969 study, Raisman, today a professor of neuroscience studying spinal cord regeneration in London, counted the synapses and described their precise locations. Provocative as his study was, it provided the kind of evidence that could not be easily discarded.

William Windle revealed impressive foresight the next year, 1970, when he called his regeneration conference in Florida, "Application of New Technology to the Enigma of Central Nervous Regeneration." He invited Geoffrey Raisman to attend that conference, even though electron microscopy and Raisman's work on plasticity did not then seem to have much to do with regeneration.

However, both the new technology and the plasticity studies would pave the way for future regeneration research.

## Good Housekeeping

Apart from electron microscopy, another revolutionary technology that would play a crucial role in regeneration appeared around this time: a series of improved methods for tracing nerve fibers. Electron microscopy and tracing techniques complement each other. Microscopes are perfect for studying nerve cells up close, but they fill the field of vision with details without capturing the general picture; tracing, on the other hand, makes it possible to visualize entire nerve pathways.

A key finding that led to one of the most powerful tracing methods arose from a study of regeneration in fish. Unlike regeneration in mammals, regeneration in fish and amphibians occurs readily and has been investigated throughout the twentieth century, albeit by a small number of scientists. Bernice Grafstein, working at the time at Rockefeller University, in New York, was studying the remarkable ability of regenerating fish nerve cells to reconnect with their appropriate targets. In order to do so, she employed a new method for labeling the nerve cells with a radioactive material, which was then carried along the axons by a cellular mechanism called axonal transport. These pioneering studies showed that fish nerve cells undergo massive changes in metabolism when they regenerate, a finding that would lead to many of the current studies on the activity of neuronal genes involved in regeneration. But the fish nerve research also revealed surprising facts about axonal transport itself.

Dr. Grafstein, a professor of neuroscience at Cornell University at this writing, recalls that when she first began to study axonal transport, it was not considered particularly important for the functioning of nerve cells. One neurochemist dismissed it as "housekeeping," a leisurely reshuffling of materials synthesized in the nucleus and required for the maintenance of the cell's skeleton and membrane. But then Grafstein and her colleagues made a seminal, unexpected discovery: Some materials moved along the axon more than a hundredfold faster than previously believed, at a speed of about 10 inches (25 centimeters) per day. It became clear that this fast transport enabled the nerve cell to perform its unique function, that of producing and transmitting the nerve impulse. Proteins and other substances manufactured by the cell body are transported to the tip of the axon, where they create or destroy molecules that allow the axon to relay the nerve impulse to the adjacent cell via the synapse.

Although one may speak of nerve cells as being "wired together" and of the "impulses" they convey, this image of the nervous system as a network of electrical cables is misleading. The axon is really a semifluid channel equipped with molecular "tracks" that can transport streams of chemicals in both directions, to and from the cell body. The nerve impulse is also an inherently chemical process, even though it generates electrical fields. Thus, the nervous system is more like a chemical factory than an electrical device.

The study of axonal transport led to an amazingly effective technique for labeling networks of live, functioning nerves. Researchers applied a radioactive material to the cell body, waited a day or two until the material traveled along the axon and into the neighboring cells, developed the film on which the radioactive material left its "fingerprints," and thus obtained a detailed map of the interconnected nerve pathways. These pathways were so clearly marked that when displayed on a projection screen, they could be seen from the back of the room. Dr. Grafstein recalls that when these nerve networks were first presented at a scientific conference in 1971, "a gasp went up from the audience."

## Sprouts of Recovery in the Spinal Cord

It was the development of radioactive tracing methods that allowed scientists investigating the spinal cord to follow up on the work of Liu and Chambers, who had reported on sprouting in animal experiments in the late 1950s. Once the tracing stemming from Dr. Grafstein's research arrived, the first study to provide new evidence that sprouting indeed occurred in the spinal cord and was indeed linked to recovery from partial injury was published. This came in 1974, and the work was done by Dr. Marion Murray, who had performed her postdoctoral studies in Dr. Grafstein's laboratory, and the late Dr. Michael Goldberger, a former student of Liu and Chambers. However, old ideas died hard. The notion of sprouting in the brain became well established following Raisman's study and those of other scientists, but in the spinal cord the sprouting of axons remained controversial for years. Many fibers in the spinal cord are longer than in the brain, and electron microscopy, which zeroes in on the points of contact between cells, was insufficient to resolve the debate—the use of tracing methods was required to obtain a fuller picture. Yet when the new techniques were applied improperly, the tracing material wandered through nerve fibers at random, creating a confusing picture. Dr. Murray, today a prominent spinal regeneration researcher working at the MCP Hahnemann University in Philadelphia remembers attending scientific conferences with a feeling that she was "on the losing team." At one conference in the late 1970s, every single study except hers and Goldberger's had suggested there was no sprouting in the spinal cord, and the session chairman delivered a half-hour diatribe against the two-decades-old report by Liu and Chambers. "It took several years out of our careers just to have to fight," Dr. Murray recalls.

Over the course of more than a decade, Drs. Murray and Goldberger conducted numerous studies—first at the University of Chicago, then at the MCP Hahnemann University—showing that in experimental animals, sprouting followed rules and occurred in parallel with the return of spinal reflexes and the use of hind limbs after a partial spinal cord injury. Eventually, evidence from these studies and those by other researchers accumulated and became accepted. And in 1988, Murray and Goldberger used new techniques to repeat the exact study of Liu and Chambers, vindicating the two plasticity pioneers 30 years after their original report.

The sprouting research prepared the ground for future regeneration studies by lifting the taboo on the idea of growth in the adult brain and spinal cord. And the tools used in this research, electron microscopy and tracing methods, later became key techniques used directly to study regeneration. Thus, the humorous adage about the importance of investigation methods in neuroscience—"the gains in brain are mainly in the stain"—also perfectly applies to the progress in regeneration research. If it weren't for the new tools, scientists would now probably still be arguing over the possibility of spinal cord regeneration, as they did over Windle's Piromen experiments. But in the early 1980s, the tracing technology, which a decade earlier had caused the scientific community to gasp, began to bring new rigor to the regeneration debate.

# 4

# 1980–1981:
# Bridges to a New Era

A picture showing specks of black paint scattered against a white background—this was the image that set in motion the current era of spinal cord regeneration. To an untrained eye, the unremarkable dots looked suspiciously like dirt, but they were a regeneration scientist's dream come true—being able to detect the origin of growing nerve fibers with unmistakable precision. It was these specks that in the early 1980s allowed scientists to demonstrate definitively that cut spinal cord fibers were capable of regrowing.

The accomplishment drew no newspaper headlines and passed virtually unnoticed outside the scientific community. Yet the specked images—which showed rat spinal neurons regenerating with the help of implanted peripheral nerve tissue—represented one of the greatest intellectual moments in the history of spinal regeneration research. Albert Aguayo and his colleagues at McGill University in Montreal asked a basic question that had never been properly answered: Are cut spinal cord fibers of adult mammals capable of growth given the proper conditions? In other words, the scientists wanted to know whether something about the spinal cord fibers *themselves* stopped them from growing, or whether something was "wrong" with their *environment*. The implications were enormous: Depending on the answer, researchers could either stop wasting time on an intractable problem or start looking for a proper environment to regenerate the spinal cord.

At that time, the handful of scientists who worked on regeneration were mostly studying situations in which regeneration was possible. Efforts focused primarily on creatures in which the central nervous system regenerates, fish and amphibians. If scientists figured out how their regeneration occurred, they could reproduce the process in injured mammals. There was also a great deal of research on the developing embryonic nervous system, which is highly adaptable even in mammals. As for adult mammals, despite William Windle's efforts, very little research was performed in brain and spinal cord regeneration; young scientists were discouraged from entering a field that was seen as a dead end. When occasional attempts were made to repair the adult spinal cord, they produced inconclusive results.

The McGill images, first published in 1980 in the journal *Nature*, changed the regeneration landscape. According to veterans of the field, they "stood the regeneration community on its ears" and "launched a new era in spinal cord research." Dr. Paul Reier, who had been studying regeneration in tadpoles in the early 1970s but had switched to the rat spinal cord by the end of the decade, says that during that period, he risked "being taken away in a straitjacket" for discussing effective spinal regeneration in mammals. The McGill research, he says, restructured the thinking on the topic: "If it hadn't been for that, I'd still look like a loony."

The tracing technology that made the specked images possible provided crucial evidence for the acceptance of the McGill findings, but other factors contributed. Most important, perhaps, was that in contrast to the past, scientists were now more willing to listen. According to Dr. Bernice Grafstein, "it's possible that Aguayo's work attracted and stimulated so many people because at last it fitted in with something; it's very hard to accept or use an idea if it doesn't fit in with what we know." The openness came from the research of the 1970s and early 1980s, which—even though it focused mainly on the brain, not the spinal cord—created a growing interest in the central nervous system's plasticity. "During that period, studies were showing how the brain continues to develop throughout its life, how it can rebuild itself, how circuitry can be reconstructed, which was a step toward seeing the brain the way we do now—as a highly plastic, dynamic organ," says Anders Björklund of the University of Lund in southern Sweden, whose laboratory Aguayo visited often for collaborative research. Professor Björklund's own work, including some of the first studies convincingly showing regeneration in the mammalian brain, had greatly stimulated the field of central nervous repair. Björklund says the impact of his early studies, as well as of Aguayo's work, stemmed from setting realistic goals: "If we had the idea to cure paralysis, this would have led to disappointment because clearly the knowledge wasn't there. With too much hype, anything other than success is seen as a failure, and then people say, 'All this is hopeless.' Instead, for us, the impetus was to get into interesting biology, which led us to take up new methods and study old questions in a different way."

Aguayo, now referred to as the "granddaddy" of spinal cord regeneration, was a perfect ambassador for the McGill results. Charismatic, witty, and influential, he tirelessly described his team's findings at endless scientific conferences and lent a much-needed aura of legitimacy to spinal regeneration research. Today Dr. Aguayo spends a good part of the year teaching neuroscience in developing countries. In 2001, he became secretary-general of the International Brain Research Organization, or IBRO, which has a strong agenda in the developing world. "I come from the Third World and I feel a need to give back," Aguayo says. He studied medicine in the late 1950s in his native Argentina, then went to Canada for a residency in neurology, after which he intended to return home, but changed his plans when a right-wing military junta seized power in Argentina in the 1960s. He shortened his name from Alberto to Albert, settled in Montreal, and became a professor of neurology at McGill University, one of the oldest and most prominent centers of neurological and neuroscience research in the world.

Aguayo specialized in disorders of peripheral nerves, but over the years he began devoting more time to research than to neurological practice. He then became head of neuroscience research at McGill's Montreal General Hospital, where an experiment curiously foreshadowing his own regeneration studies had taken place at the beginning of the twentieth century. One of Montreal General's leading surgeons tried to cure a paralyzed sailor by replacing the lower part of his damaged spinal cord with the spinal cord of a large dog, then stimulating the sailor's paralyzed muscles with electrical currents for a few hours every day. The man regained no function and died of infection, but the surgeon argued in a 1905 scientific paper that a limited regeneration of axons had taken place in his patient's spinal cord.

## Controversial Inspiration

The experiments carried out by Aguayo and his team were a modern replay of studies performed at the turn of the century by Francisco Tello in Santiago Ramón y Cajal's laboratory, then repeated by other researchers. In those studies, scientists created a gap in a nerve belonging to the central nervous system and bridged this gap with a piece of a peripheral nerve, known for its ability to regenerate.

Regeneration of peripheral nerves is a slow but sure process. When a nerve is cut in two, only the part that is attached to the cell body survives. The severed stump quickly deteriorates and dies, and its protective sheath is left lying like an empty sleeve. However, the sheath is lined with cells secreting growth-promoting materials that can give new life to the nerve. The surviving piece of the nerve begins to grow and crawls into the empty "sleeve," spurred on by the secretion from the sheath's inner lining. After a few weeks or months, the nerve, which grows at about the same rate as hair, approximately 0.5 to 1 millimeter per day, reaches its previous length, and restores function to the muscle, at least to some degree. Thus, when a surgeon sutures a cut peripheral nerve, while reattaching a severed finger for example, the goal is not to put together the two pieces of the nerve, but to create a path for the surviving piece to regrow. The rationale behind researchers' peripheral nerve transplants was to create the same path for a nerve of the brain or spinal cord. Perhaps the central nervous fibers could be coaxed into regenerating if provided with such a friendly, growth-stimulating "environment."

Tello had shown successful regrowth of brain fibers into the transplants of peripheral nerve, but whenever scientists tried to repeat these results, they ran into the difficulty that had plagued all early regeneration experiments: They could never prove that the growing fibers originated in the central nervous system. In the early 1940s, the prominent British anatomist Wilfrid Le Gros Clark repeated the Spanish study as part of research aimed at ultimately treating wartime injuries, but—being "very British and very critical," says Aguayo—proclaimed that the Spaniards had made a mistake. The growth of new fibers in the brain, Le Gros Clark said, was nothing but "contamination" from regeneration of the neighboring peripheral fibers in the scalp, while the brain fibers "remained quite inactive." In the absence of proper technology it was impossible to resolve the argument, and the negative opinion prevailed. Only in the early 1980s, thanks to the McGill experiments, did scientists prove that Tello and Cajal had been right after all.

The impetus for the regeneration studies at McGill, however, did not come from the historic Spanish experiments (even though Aguayo, perhaps inspirited by ethnic solidarity and coming from a country where neuroscience was strongly influenced by the school of Ramón y Cajal, is a self-proclaimed Cajal buff who collects books and anecdotes about the revered Spanish scientist and keeps several portraits of Cajal in his office). It would have made for a nice story had Aguayo's regeneration experiments been directly inspired by those performed in his hero's lab, but the reality was more

complicated. The history of spinal regeneration research is pep-
pered with daring—or, depending on the point of view, daredevil—
figures who came up with solutions aimed at curing paralysis over-
night. The immediate trigger for Aguayo's studies was provided by
one such man, a Taiwanese-born neurosurgeon named Carl Kao.

Dr. Kao was a controversial kind of pioneer. He had moved to
the United States in the mid-1960s and started off as a well-known
and respected figure in the medical and scientific world after doing
a residency with Leslie Freeman, a famous Indiana surgeon who
treated many spinally injured people and experimented a great deal
with spinal cord repair in rats. Kao won an award for an innovative
spinal cord transplantation technique and coedited an important
book on spinal cord reconstruction, but his claims of speedy cures
for cord injuries drew criticism from his peers. In the late 1970s, he
transplanted pieces of peripheral nerves into the injured spinal
cords of paralyzed dogs, showed that the dogs could move their
hind legs, and argued that the animals had recovered. Other scientists were unconvinced—the dogs' condition may have improved
anyway, or perhaps they were just moving their legs in a reflexive,
involuntary way. Kao provided electron microscope images of growing fibers, but as usual, it was unclear whether the fibers belonged
to the spinal cord. Nonetheless, Kao went on to perform his surgery on people with spinal cord injuries. Much of the scientific community was scandalized: Promising to cure humans with an unproven approach was widely viewed as charlatanism. "The trouble
was that anybody who came along and argued, 'I can cure these
spinally injured dogs or people,' was immediately suspect, because
experience said this is not something that just happens right off the
bat," Dr. Grafstein says.

Dr. Kao today has an office in the Washington, D.C., area, but
he operates on patients in South America, inviting spinally injured
people to travel there from the United States to have his surgery.
Kao says he moved his operations overseas because insurance in the
United States would not pay for them, making it impossible for
most people to cover the high hospital fees, but in any event, most
U.S. hospitals would probably not allow him to perform his still

unreplicated procedure: Innovative surgeries go into hospital use after being endorsed by committees relying greatly on peer review. His operations include exposing the spinal cord and replacing the injury site with transplants of peripheral nerve and a patch of blood-vessel-rich abdominal tissue called omentum. According to unconfirmed reports, some people regain partial use of their formerly paralyzed bodies after Kao's treatment, but these results do not necessarily validate Kao's claims of regeneration. Surgery that frees the spinal cord from the tethering to surrounding membranes, without transplants of any sort, has long been known to improve the condition of some people, particularly if performed by a vastly experienced surgeon like Kao. Since Dr. Kao has not published his results with humans in the major peer-reviewed scientific journals, his approach remains controversial and it remains unknown whether it works and how, even though according to him, by the year 2000 he had operated on some 500 people. Back in the late 1970s, Dr. Kao's claims antagonized many scientists, but according to old-timers, he deserves a great deal of credit for bringing a high profile to the field. He stirred up debate and interest that led to other experiments, including those at McGill.

Aguayo decided to repeat the transplantation experiments on animals, working with a younger colleague at McGill, the talented, unassuming neurosurgeon Peter Richardson. The two had known each other since Richardson was a teenager: When Aguayo first came to Canada, he was trained by Peter's father, Clifford Richardson, considered one of the best neurologists in the area. Later, Aguayo became a mentor to the younger Dr. Richardson when Peter obtained a position at McGill that allowed him to combine clinical neurosurgery with research. The two physician-researchers had heard Kao speak at a conference, were intrigued by his ambiguous findings, and together devised an elegant method to resolve the controversy that had surrounded peripheral nerve transplants since the beginning of the twentieth century. "We did exactly the same experiment as many people before us but with two things—a new technology and a little more conviction," Aguayo says.

## The Horseradish Solution

The technology Aguayo and Richardson used stemmed from the revolution that had occurred in the tracing of nerve fibers during the preceding decade. By then, scientists had even learned to trace the nerve pathways backward—from the ending of the axon to the cell body. They found that one tracing material, an enzyme derived from the horseradish plant, was particularly suitable for this purpose; it was readily taken up by the axon and transported back to the cell body, as if sucked up by a vacuum cleaner. The enzyme reacted with a special dye, painting the course of the nerve pathway and marking the cell body with a dark dot.

Aguayo and Richardson had one of those brilliant ideas that in retrospect—after they have proved successful—seem obvious and simple; they decided to apply the tracer to regenerating fibers. They removed a 1-centimeter piece of a rat's spinal cord and bridged the gap with a piece of nerve from the rat's leg. Once the fibers would grow through the "bridge," the scientists would inject the tracer to find out where the growth was coming from. "It's amazing that other people didn't do it at the same time," Richardson says.

While waiting three to four months for the rats' spinal cords to grow, the scientists were confronted with the challenge of caring for the paralyzed animals. Today laboratories working on spinal cord regeneration function like minihospitals with trained personnel, but in the early 1980s the task found the McGill researchers unprepared. The rats needed close attention and had to have their bladders and bowels emptied manually on a regular basis. To provide the animals with round-the-clock care, Dr. Richardson enlisted the help of hospital nurses who worked with his human patients. "I don't think they've ever forgiven me," he says.

The experiment culminated with the scientists injecting the horseradish tracer into the peripheral nerve "bridges" inside the rats' spinal cords, which now contained new, freshly grown fibers about 1 centimeter long. The tracer traveled through the fibers and "reported back" about their origins, the cell bodies. These were regenerated spinal cord fibers—and not any others—because their cell bodies sat *inside* the cord! A slice of spinal cord tissue just below the bridge was dotted with dark specks, cell bodies filled with the horseradish tracer. Had the growing fibers come from elsewhere, the tracer would have had no way of getting into the cells inside the spinal cord.

Even though the 1980 study appeared in *Nature*, the centuries-old skepticism regarding spinal cord regeneration persisted. "Basically, the message from the scientific community was, 'This sounds very interesting but it could still be an artifact,'" Aguayo recalls. With a relatively short bridge, there was the concern that despite all the precautions (such as spreading vaseline around the graft to isolate it from surrounding tissue), the horseradish dye might have somehow diffused into the spinal cord rather than traveling through the growing fibers. To produce an indisputable proof, a longer nerve bridge had to be built.

Aguayo gave the assignment to build such a bridge to a new postdoctoral fellow in his lab, Samuel David. David had worked in rehabilitation of people with paralysis in his native Bombay, and continued this work after moving to Canada. He still remembers some of his patients: a young man who was startled by a deer and flew off his motorcycle while crossing the Rocky Mountains; a teenage girl who sat with her feet up in the back of a car during a family holiday in Spain when the car was smashed in a traffic accident. Moved by these cases, David looked into the possibility of studying spinal cord regeneration when he started working toward his Ph.D. at the University of Manitoba in 1974, but his research supervisor suggested he pick another topic for his doctoral thesis. "You have your entire life in front of you, and this field is going nowhere at present," the supervisor said. David did his thesis on the development of the nervous system, but toward the end of his studies he heard Dr. Carl Kao report on peripheral nerve transplants into the spinal cord at a scientific conference. "I think he overinterpreted his results, but I remember being very excited about them," David says. When he came to do his postdoctoral studies in Dr. Aguayo's laboratory at McGill, he was thrilled to discover that in one of the lab's projects, Aguayo and Richardson were applying tracing methods to peripheral nerve bridges.

The scientists came up with an unorthodox solution to the long

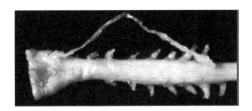

Diagram and photo of the handle-like nerve "bridge" attached to the spinal cord of a rat. Even though the spinal cord remained intact, the peripheral nerve bridge prompted some of the cord's fibers to grow. Scientists David and Aguayo, who had experimented on mice before moving on to rats, noted that the fibers grew through the bridge in both directions over a great distance. (Courtesy of Carol Donner.)

"bridge" problem. Because removing a very long chunk of the spinal cord would have jeopardized the life of the experimental animal, David created a bridge, or rather a bypass, *outside* the animal's intact spinal cord. He removed a 3.5-centimeter-long piece of nerve from a rat's leg and attached it under the skin on the animal's back, with both ends anchored in the spinal cord, one end at the level of the neck and the other around the area of the chest. On a side-view diagram, the bridge looked like a handle on top of the rat's spinal cord. As an extra benefit, this approach eliminated the need to care for paralyzed animals: The healthy rats spent six to seven months running around the cage unaware of the "handles" on their backs. "As soon as the spinal cord fibers saw that peripheral graft they just took off," David says. In these experiments, the scientists were again able to trace the fibers all the way back to their origin, the central nervous tissue in different parts of the brain and spinal cord, with the help of the horseradish tracing technique. The report on the study, published in 1981 in *Science*, left no doubt that these were spinal cord fibers growing.

The "handles" were a caricature of growth, Aguayo says. Like a caricature that exaggerates a feature, such as a big nose, the fibers in these experiments performed an "exaggerated" act of regrowth: They traveled along the roundabout route and grew two to three times the distance that they would normally have needed to go to fill in the gap. "We were exaggerating in order to make a point," Aguayo says. He and his colleagues placed the long bridges all over the brain and spinal cord, including the optic nerve, the cortex, and the subcortex, and almost everywhere, the fibers grew long distances. Scientists at other institutions repeated the studies and obtained similar results, confirming the McGill findings. In each case the number of regenerating cells was relatively small, usually no more than 10 percent of all fibers, for reasons that are still unclear: Is only a small subset of nerve cells capable of regrowth? Or do the others simply require different conditions? Scientists are still exploring these questions, but the 10 percent sufficed to show that spinal cord regeneration was a legitimate scientific pursuit.

# 5

# The Unyielding Spirit

The McGill experiments generated excitement, but they also created frustration. Everybody was expecting that peripheral "bridges" would help paralyzed animals walk, but they did not. The growing fibers stopped dead in their tracks when they reached the end of the bridge; they refused to reenter the other side of the spinal cord. As Albert Aguayo put it, "nature isn't saying 'yes' so quickly." Without recovery of function, the growth of nerves threatened to remain a purely intellectual feat.

Aguayo began to address this problem together with his close McGill collaborator, Professor Garth Bray, a neurologist and an expert on electron microscopy. They had merged their laboratories early on in their collaboration and worked together on many, although not all, research projects. The two scientists are a study in contrasts: Aguayo, with his bushy mustache, warm brown eyes, and a love for opera and tango, is ebulliently Latin in both appearance and character; Bray, a tall, pale northerner, who grew up in the chilly province of Manitoba in central Canada, has a reticent, understated manner. Aguayo is a dreamer, good at articulating grandiose concepts and generating enthusiasm; Bray, described by his colleagues as "a solid citizen type" and "salt of the earth," has over the years served as the laboratory's reality check—he was merciless in questioning all new findings and insisting on rigorous criteria for every study. Aguayo held prominent administrative positions at McGill and nationally, but he and Bray jointly ran a busy, multinational laboratory at Montreal General Hospital for more than 30 years.

A nerve cell is of no use unless its fiber effectively connects and communicates with other cells. Aguayo and Bray set out to explore a question that is key for recovery of function and that today remains the central unknown in spinal regeneration research: Can a nerve fiber that has regenerated over a long distance make working connections, or synapses, with the rest of the central nervous system? The question of new connections is extremely difficult to resolve in the spinal cord, where neurons communicate with many different nerve networks. Therefore, Aguayo and Bray focused on a more accessible, uniform, and simple bundle of central nervous fibers, the optic nerve. This nerve is the eye's only connection with the brain; it transmits information from light-sensitive receptor cells in the retina to appropriate brain regions.

Because nobody in the laboratory had much experience with the optic nerve, Aguayo enlisted the help of the MIT-trained Chinese researcher Dr. Kwok-Fai So, who specialized in studying the visual system. During his relatively brief, six-month stay in Montreal, Dr. So invented the surgical procedure for studying optic nerve regeneration. Employing microsurgery and a needle so tiny that if dropped on the floor it could never be found again, he cut the optic nerve fibers in the retina and placed a piece of peripheral nerve right inside the eyeball. Using this method as a basis, the skilled Spanish surgeon Manuel Vidal Sanz, who joined the laboratory at about the same time and stayed for nine years ("the first Spaniard to arrive," Aguayo enthuses), invented another procedure for studying recovery of function after regeneration: He entirely replaced a rat's optic nerve with a graft of peripheral nerve from the animal's leg.

When the results of the grafting procedure were analyzed three months later, the scientists found that optic nerve fibers had regenerated along the graft. But it took two years of such studies to find the first synapses between the regenerated fibers and the brain region with which the optic nerve normally makes connections. The scientists labeled the nerve fibers and, using an electron microscope, scanned the brain for labeled synapses, much like treasure hunters scanning the ocean floor for a sunken ship. They cut a section of the brain, made a grid, and systematically went from left to right— boring, boring, boring. "It was a very frustrating time," Dr. Bray recalls. "And a very exciting time. Finally finding large numbers of synapses formed by regenerated fibers was truly the emotional highlight of my research career." The results, first reported in 1987, were rewarding. The synapses looked normal in size and shape, they formed in the appropriate layer of the brain, and their pattern of distribution was the same as in normal, unoperated animals. Moreover, the connections had remained in place for a full year and a half. Everything suggested that the regenerated fibers had formed robust circuits.

In parallel, a neighboring laboratory at Montreal General Hospital became involved in this research in order to determine whether the regenerated optic nerve behaved normally—in other words, whether it was capable of doing what nature had intended it to do, relay visual information to the brain. The laboratory was headed by Dr. Michael Rasminsky, another physician-researcher who specializes in neurophysiology, the electrical study of the nervous system. Like Aguayo and Bray, Rasminsky saw patients at Montreal General, and he led animal studies on conduction of nerve impulses in a laboratory down the hall. The optic nerve study was assigned to Susan Keirstead, a freshly minted Ph.D. from Queen's University, in Kingston, Ontario, who had studied the spinal cord but had a special interest in vision that had been aroused by her sister's and brother's congenital visual handicaps.

Aided by Manuel Vidal Sanz and the neurosurgeon David Carter, Keirstead and Rasminsky tested the functioning of the newly formed connections made by the regenerated optic nerve in rats and hamsters. Working in a dark laboratory, they flashed a bright light onto a screen positioned in front of the experimental animal and monitored activity in the animal's brain with electrodes. The only neurons that respond to light are those in the retina, so if the brain received a signal about the flashes, this would be a clear indication that the connections formed by the regenerated optic nerve were working.

Activity of a neuron is often referred to as the neuron's "firing" because each signal generates electrical changes that, when picked up by an electrode connected to a sound amplifier, produce a sharp, crackling sound. In their experiments, while exposing the eyes of laboratory animals to light, Keirstead and Rasminsky placed electrodes into an animal's brain to check whether its optic fibers reported to the brain about the flash. (The experiments were performed on anesthesized animals, who felt no pain.) Probing with electrodes the tiny brain surface where connections were thought to be located was a lengthy and tedious process. For every animal, scores of probes were necessary, each taking up to half an hour or more. Every fraction of a millimeter is an enormous distance when it comes to nerve cells in the brain. "You move half a millimeter away, and you're bombing Paris instead of Berlin," Rasminsky says. After about a year the scientists still had no results. At that time a new team member from Japan, with expertise in the neurophysiology of the visual system, Yutaka Fukuda, joined the project.

Having failed seven times to obtain a professorship at various medical schools, Dr. Fukuda was at a critical stage in his career. He brought with him a good-luck charm—a red, humpty-dumpty-like doll, a *daruma*, named after a legendary Buddhist monk; the doll symbolizes an unyielding spirit and the belief that after seven failures, one could still succeed on the eighth try. The Japanese often buy a new *daruma* before embarking on a difficult task, such as an election campaign or a university entrance exam; they paint a black

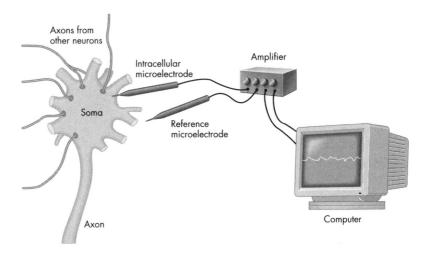

The apparatus used for recording a neuron's response to stimulation. (From *Biological Psychology*, 7th Edition, by J.W. Kalat, 2001. With permission of Wadsworth Group, a division of Thomson Learning.)

dot in one of the doll's eyes and pray for success. When they succeed in the task, the *daruma*'s second eye is painted black, as a sign of appreciation.

The experiments continued for about three more months with the one-eyed *daruma* watching from atop one of the monitors, until late one night, after flashing a light into the animal's eye, the scientists for the first time heard a sound of a firing neuron. In the silence of the empty lab, the telltale *khrrkh-khrrkh*, which resembles static electricity on a telephone line, was music to their ears. The connection was working! The regenerated nerve was sending a message about the flashed light to the brain, and the brain was getting the message. "I remember very clearly walking home at 4 a.m. with a feeling of tremendous elation," Dr. Rasminsky says. "So much time had been invested in this study, and it wasn't by any means clear that it was going to work." In subsequent weeks the finding was confirmed, the *daruma*'s second eye was painted black, and the experiment's results were sent to be published in *Science*. (Shortly afterward, Dr. Fukuda applied for a position for the eighth time and obtained a professorship at Osaka University Medical School, where he was still working at the time of this writing.)

Rasminsky is at a loss to explain why the working connection had been so difficult to find. "Sue Keirstead and I were doing this experiment night after night just getting nothing, and it wasn't clear why. There is a wonderful quote from Sir William Henry Bragg, a Nobel laureate in physics, who said toward the end of his career: 'After a year's research, one realizes that it could have been done in a week.' In retrospect, we felt exactly the same way—before getting it right we had spent a tremendous amount of time just spinning our wheels." Dr. Aguayo makes an interesting point about the electrical recording study and the persistence that it required: The electrodes for recording neuronal signals had existed long before sophisticated tracing methods appeared, which means that some of the experiments in Rasminsky's laboratory could have been done much earlier than they were, probably making a convincing argument in favor of central nervous regeneration years, if not decades, before the 1980s. However, before the Richardson and David tracing studies, there was not enough confidence in the possibility of regeneration to justify the labor-intensive electrical recording experiments.

Did the restored optic nerve connection mean that the animals could see again? Aguayo, who was invariably asked this question at scientific meetings, had an arsenal of humorous replies (for example, "I don't think they can read the papers, but they never could"). Rodents rely a great deal on smell and touch and may appear to behave normally even when they cannot see. The serious answer was that the animals' sight had probably not been restored. However, the experiments provided the first indication that regenerating nerve fibers could make new, working connections within the adult central nervous system. Even if the new connections work, do they function properly? And if so, will such connections form after spinal cord injury in humans? Scientists are still struggling with these questions, but the studies led by Aguayo and Bray were the first step toward providing an answer. Former members of their Montreal team would later conduct independent follow-up studies in their respective countries, showing that a regenerated optic nerve can perform several clearly defined tasks. Yutaka Fukuda and his Japanese colleagues showed that hamsters can learn to discriminate between light and darkness while relying on regenerated optic fibers. Kwok-Fai So, now director of neuroscience research at the University of Hong Kong, in collaboration with American researchers, demonstrated that regenerated optic nerve fibers allow hamsters to follow visual cues. And Manuel Vidal Sanz, working with British researchers, showed that a regenerated optic nerve of a rat can trigger the contraction of the pupil in response to bright light. "We cannot conclude from these results that the entire spinal cord is going to regenerate and reestablish appropriate connections," says Vidal Sanz, now a professor of experimental ophthalmology at the University of Murcia, in Spain. "However, we showed that, under certain conditions, a specific group of fibers can reestablish connections and restore function, and this is encouraging."

After the McGill findings, there was no more turning back for spinal cord repair. Inspired by the knowledge that spinal cord fibers could regrow, other scientists started looking for clues to the most successful regeneration.

# 6

# A Schwann Cell Chauvinist

T he white-enamel surfaces in the lab of Richard Bunge at Washington University School of Medicine in St. Louis were tattooed with black felt-pen graffiti. The scientists used the enamel coverings of equipment to display their own aphorisms, other people's quotes, and whatever else sparked their imaginations. On the front panel of one air-filtering hood, someone had inscribed a phrase that captured the creative spirit of the lab: "Nothing is ever true in biology." One biological "truth" that Bunge encouraged his team to question was the inability of the central nervous system to regenerate.

Dr. Bunge, an eminent neuroscientist and a warm, gregarious man, who had been trained as a physician before going into research, was not only widely respected in the scientific community but also immensely well liked. He was an expert on myelin, the protective sheath of nerves, and a world leader in the study of Schwann cells, which make myelin for peripheral nerves. In the early 1980s, stimulated by studies of his long-time collaborator Albert Aguayo, he decided to apply his Schwann cell expertise to spinal cord regeneration.

The decision to focus on Schwann cells was slightly off the beaten track for a neuroscientist. Schwann cells belong to the nervous system's supporting cells, collectively known as the glia (from the Greek word for "glue"). Originally, these cells were regarded as nothing but glue holding the neurons together, but with time the glias' role emerged as much more important than that. For many years, starting in the late 1970s, Dr. Bunge led informal discussions about Schwann cells within the framework of conferences on peripheral nerves. Participants in these evening get-togethers once called themselves Schwann cell chauvinists, jokingly defending their decision to focus on the "second-class" glia rather than the more glamorous neurons. At one conference in Berlin, after visiting what was then the Communist part of the city, they invented a slogan, "Schwann cell chauvinists of the world, unite!" At these gatherings, and at many larger, more formal scientific conferences, Bunge, a tall, imposing man with a disarming smile, often revealed his gift for bringing the discussion into focus with a sharp remark. Once at a symposium on pain, when a speaker tired the audience with exceedingly lengthy comments, Bunge asked him to come to the point, adding, "We are here to relieve pain, not to inflict it."

During nearly four decades, Dr. Bunge worked side by side with his wife, Mary, whom he met when they were both students at the University of Wisconsin School of Medicine. They were one of the most famous husband-and-wife teams in neuroscience, thanks to a series of landmark studies on myelin and myelin-related diseases. Their personalities complemented each other: Richard, nicknamed by his team "The Captain," made his presence known wherever he went; Mary, a tall, slim, elegantly dressed woman, is quiet and reserved. Richard was a visionary, a big-picture man; Mary has a strong eye for detail. Her lifelong interest in art drew her to electron microscopy, in which the quality of the image can often be judged by its esthetic value. In 1960, the Bunges moved from Wisconsin to Columbia University in New York City, and in their subsequent career moves they followed Richard's appointments, while Mary would obtain a position in the same department. They created a supportive, almost family atmosphere in the laboratory and, apart from spending time with their two sons and being active in the Presbyterian Church, devoted all their energy to science, blurring the line between recreation and research. When Richard Bunge built a wooden table for a microscope in his laboratory, he decorated it with diamond and half-moon shapes symbolizing Schwann cells clinging to a nerve fiber. When a scientist was to leave their lab, Richard and Mary threw a going-away party in their home, for which Richard composed a farewell song that he performed while accompanying himself on a guitar. The song inevitably contained puns and rhymes about Schwann cells ("If you've ever seen a Schwann cell/With an axon in its grasp/And a basal lamina all around,/You've seen a love 'twil last").

Richard Bunge always referred to the Schwann cell as "the cell of Schwann," as if giving tribute to the cell's discoverer, the German scientist Theodor Schwann, every time he used the word. Schwann was 29 when, in 1839, he published a book said to be second only to Darwin's later *Origin of Species* in its impact on biology. It was a treatise on the "similarity in the structure and growth of animals and plants," which formulated one of the most fundamental biological concepts: that all living things are made of cells. Schwann was deeply religious and, before publishing the manuscript, he submitted it for approval to his archbishop to make sure it was not contrary to the doctrine of the Catholic Church. He usually shares credit for the cell theory with another German scientist, Matthias Schleiden, who had formulated the same idea for plants. Far from being hailed as a prodigy, Schwann was ridiculed by Germany's scientific establishment for another concept he proposed while still in his twenties: that fermentation was the work of yeast, a live organism. This idea, later proved to be true, was received with

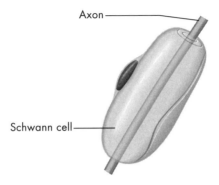

Axon

Schwann cell

A Schwann cell wrapped around a nerve fiber, where it deposits multiple layers of insulating myelin. (From *Biological Psychology*, 7th edition, by J.W. Kalat, 2001. With permission of Wadsworth Group, a division of Thomson Learning.)

such hostility that Schwann was forced to move abroad, where he led, according to one biographer, a "solitary existence darkened by episodes of depression and anxiety."

When Schwann first identified the cell that would later bear his name, he called it the "primary nerve cell." It took more than 100 years for the function of the Schwann cell in the nervous system to be clarified. With the advent of electron microscopes, scientists found that during the development of an embryo, a Schwann cell repeatedly wraps its surface membrane around the peripheral nerve fiber in a jelly-roll fashion, to deposit multiple layers of myelin. When an adult peripheral nerve is injured, Schwann cells begin to divide, provide nourishment and guidance for the regenerating axon, and restore the protective myelin wrapping. Bunge's idea was that in the spinal cord, a bridge made of Schwann cells would do a better job at promoting regeneration than a transplanted piece of peripheral nerve. He also hoped that Schwann cells would restore the myelin sheath to injured nerves.

## The "Jumping" Conduction

Without the myelin wrapping, which looks white because of the fat it contains, the spinal cord cannot do its job. To understand why, consider the difference in the way a nerve impulse travels along unmyelinated and myelinated axons.

In unmyelinated axons, the chemical and electrical changes that constitute the nerve impulse spread through the membrane. The spread is a time-consuming chain reaction that propels the impulse relatively slowly, at approximately 3 to 33 feet (1 to 10 meters) per second. At this speed, for example, it would take the brain of a giraffe about a second, an unacceptably long time with a lion in the vicinity, to send a signal to the foot. (Still, some shorter and thinner axons in the brain and spinal cord are covered by little or no myelin. This is true, for example, of nerve fibers that transmit the feeling of pain. If you put a hand on a hot plate, the sensation of touch, which travels along myelinated fibers, comes first, and the pain follows shortly afterward.)

The thicker and longer axons in the nervous system of vertebrates are usually covered with myelin, and they conduct nerve impulses differently and much faster. The myelin sheath is an excellent insulator that does not allow for chemical and electrical changes to be triggered in the membrane. However, the sheath is interrupted at intervals of approximately 1 millimeter by tiny spaces with no myelin, called nodes. In myelinated axons, the membrane activity "jumps" from one node to another, while in between the nodes, the impulse takes the form of a current of charged particles inside the axon that spreads almost instantly, at the speed of electricity in a conducting fluid. The jumping of the impulse from one node to another is called saltatory conduction, from the Latin word *saltare*, meaning "to jump." One neuroscience professor draws an analogy between saltatory conduction and an Olympic sprinter who covers a given distance in large strides; in contrast, conduction along unmyelinated axons is equivalent to covering the same distance while walking heel to toe. Thanks to the "jumping" conduction, electrical signals can travel down myelinated nerve fibers at a dizzying speed of about 330 feet (100 meters) a second. This is vastly slower than conduction of electricity in a wire but sufficient for coordination of rapid responses even in such large creatures as giraffes, whales, and humans. (Human babies are somewhat clumsy partly because their nerve fibers are not fully myelinated until the age of three.)

Loss of myelin as a result of injury or disease can not only slow down but actually interrupt the conduction of nerve impulses. Signals are halted or blocked completely by the denuded parts of the nerve, just as the burning of a gunpowder fuse fades in its moistened segments. And when muscles receive no timely signals, disruption of movement or even paralysis can occur. In some diseases, such as multiple sclerosis, loss or malfunction of myelin is the major problem. In spinal cord injury, too, some of the nerve fibers, even if they are not cut, cease to conduct impulses because they lose their myelin.

Richard and Mary Bunge were the first to demonstrate in the early 1960s that in the adult mammalian central nervous system, myelin can be broken down and then reformed—a radically new idea at the time. This discovery was made with an electron microscope, in those days still a relatively primitive instrument whose lenses often had to be hammered into place. Another seminal finding the couple made with the same microscope concerned the way myelin is formed in the brain and spinal cord. At the time, it was already known that Schwann cells make myelin for peripheral nerves. As for the central nervous system, the prime candidate for myelin manufacture was the oligodendrocyte, an octopus-shaped glial cell, but its role in myelination was difficult to prove. Oligodendrocytes have a less intimate relationship with nerve fibers than do Schwann cells: The main part of the cell does not wrap around an axon, as a Schwann cell does; rather, the myelin is deposited at the distant ends of the "octopus" arms, and as a result, it is more difficult to detect which cell is responsible for myelin formation. In 1962, the Bunges published the first images of an oligodendrocyte's tentacles depositing myelin in a developing mammalian spinal cord—proof that these cells were indeed responsible for covering the central nervous axons with myelin. Seeing the myelin form, Mary Bunge recalls, "was one of those moments that makes all the work worthwhile." It is now known that a single oligodendrocyte can place myelin coating on as many as 40 fibers.

Upon moving in 1970 to Washington University, one of the hubs of American neuroscience, Richard and Mary devoted much of their time to exploring fundamental questions about "the cell of Schwann," using tissue culture methods developed by their team member Dr. Patrick Wood: How does the cell communicate with the nerve fibers and what exactly can it do for these fibers? What encourages the Schwann cell to form myelin? Among the laboratory's numerous discoveries, Drs. Wood and Richard Bunge revealed that an injured peripheral nerve appeared to release a chemical "Help!" signal that causes Schwann cells to divide, but at the time the nature of this signal was unknown.

In 1989, Richard Bunge accepted an offer to become scientific director of the recently established Miami Project to Cure Paralysis, whose upbeat name was inspired by the success of the Manhattan and Apollo projects. Located on the sunny, subtropical campus of the University of Miami School of Medicine, the Project, said to be the largest research center in the world devoted entirely to spinal cord injury, has for a logo a figure of a person getting out of a wheelchair. Here Bunge launched a large-scale study into the pathological changes that occur in spinal cord tissue after injury. One important finding focused on myelin; it was already known that loss of myelin occurs in spinally injured experimental animals, but Bunge showed this also happens in a damaged human spinal cord. It is now believed that some people with only partially damaged spinal cords remain paralyzed because their surviving nerve fibers have been stripped of myelin, which prevents these fibers from conducting impulses; some function may be restored after injury if these spared fibers are remyelinated. Moreover, many regenerating fibers are likely to function only if—after regrowing—they become covered by myelin.

## The Bridges of Dade County

In Miami, Richard Bunge initiated a research program aimed at using Schwann cells in spinal cord regeneration of humans. "I don't want to go through years of research without knowing how to apply this therapy to people," he said. At the time, his team already knew how to grow large quantities of rat Schwann cells, but the same methods did not work for human cells. No matter what they did, the researchers could not get human Schwann cells to divide in large numbers in a laboratory dish.

The solution to growing the human Schwann cells would come from an unusual, decade-long pursuit, a story in itself in which several teams in three independent lines of research, unbeknownst to one another and separated by thousands of miles, hunted for what turned out to be the same molecule. In 1992, in a stunning convergence of different research fields, the object of their separate pursuits revealed itself as a versatile growth-stimulating molecule, now known as neuregulin. The molecule is involved in the division of Schwann cells, in nerve–muscle communication, and in the aberrant growth of cancer cells. ("It's always nice to have company," one of the team leaders said about having learned that several dozen other people had been searching for the same molecule as the one his own laboratory pursued for more than a decade.) Richard Bunge learned about the many faces of neuregulin at a scientific conference and obtained it from a company, which was, ironically, using it for a purpose directly opposed to his own: In search of a cancer drug, the company was blocking neuregulin in order to prevent the division of cancer cells. Bunge used neuregulin as a "Help!" signal to stimulate the division of human Schwann cells.

The research model the Bunges chose for their regeneration studies was a complete cut of the spinal cord, a choice affected both by research and geographic considerations: About one-third of the human spinal cords in the Miami Project's postmortem tissue bank were completely severed by the injury, in part reflecting the fact that gunshot wounds, which can fully cut the cord, are a part of the Miami scene. The scientists created a tiny gap in the rat spinal cord, which ensured that no fibers in the cord had been spared and which mimicked the complete human injury. They then assembled half-inch long cables, each made of some 6 million rat Schwann cells, and packed every cable into a polymer tube that filled the gap and provided a guidance channel for thousands of regenerating axons. Richard Bunge referred to the cell-loaded tubes as "the bridges of Dade County," after the Florida county that is home to the Miami Project.

Thanks to neuregulin, Miami Project scientists can now produce human Schwann cells in sufficient quantities to bridge a gap in the adult human spinal cord. After removing a small piece of peripheral nerve from a person's leg, within four to six weeks the scientists can produce enough cells to make a cable that is nearly half an inch (more than 1 centimeter) wide and about 18 feet (5.5 meters) long. After Richard Bunge died in 1996, Dr. Mary Bunge carried on their studies. She and her colleagues, while still working with rats, are conducting experiments aimed at realizing Richard Bunge's idea: using Schwann cells to regenerate the human spinal cord.

Schwann cell cables have revealed themselves as potent promoters of fiber growth in the spinal cord. When placed as a bridge between two spinal cord stumps, the cells stimulate axons from both stumps to grow long distances into the bridge. Moreover, Schwann cells are good at myelinating axons, both the ones that have regenerated and those that have survived but lost their myelin as a result of injury. (Myelination of denuded fibers is easier to achieve than regeneration, and it is an important part of spinal cord repair strategies; in fact, myelination *alone* can probably help restore certain function in people with partial spinal cord injuries.) The Schwann cells are known to produce a host of molecules useful for regeneration: growth factors, as well as support molecules that create a favorable surface on which the axons can grow.

Schwann cells provide for greater flexibility in nerve repair than transplants of whole peripheral nerves. Thus, scientists at the Miami Project have learned to make the cells "smarter": Using genetic engineering, they endow the Schwann cells with genes for proteins that the cells may make in insufficient amounts, such as growth factors particularly valuable for regeneration in the central nervous system. Besides, unlike transplants of whole nerves, Schwann cells can be more easily incorporated into combination therapies that take advantage of different regeneration approaches. However, scientists at the Miami Project have found that Schwann cells alone do not stimulate the axons to grow beyond the bridge, and therefore they are testing other strategies, in combination with bridges, to achieve this goal.

While the Schwann cell experiments were going on, the field of spinal cord regeneration had grown substantially. A new direction of research had appeared: In addition to making damaged axons regrow, scientists wanted to understand *why* effective regeneration fails to happen in the mammalian spinal cord.

# 7

# The Little Antibody
# That Could

In early regeneration studies, the biggest unknown was the very possibility of regeneration in the adult mammalian spinal cord. The studies by Aguayo and others resolved that question, but once resolved, this issue gave way to a new quandary. The problem no longer was whether spinal cord fibers *can* regrow, but rather, why *don't* they? Nature has obviously made a deliberate effort to prevent regeneration in the central nervous system, but what barriers has it set up to achieve its goal? The first such barrier, a molecular weapon that actively fights the growth of nerve fibers in the adult spinal cord, was unmasked by the Swiss scientist Martin Schwab.

The idea that the spinal cord contains inhibitors to growth had been raised earlier, but Schwab designed concrete experiments to prove it and stood by his findings until they turned into a textbook concept. It became clear that the spinal cord is not a passive flap of tissue waiting to be fixed. Any effort to promote growth in the cord would have to overcome the invisible molecular inhibition. In fact, for a while it seemed that the regeneration puzzle had been solved. If there was one mighty inhibitor of growth in the cord, all researchers had to do was to find it and clear it away. "The impression Schwab had made on scientists was tremendous," recalls the chief executive of the International Spinal Research Trust. "Some of them were in despair, like 'He's done it and I haven't.' On the other hand, there was an enormous amount of enthusiasm—yes, now we can do it, we can regenerate the spinal cord."

Today Professor Schwab heads a team of some 25 researchers, students, and technicians at the University of Zurich, Switzerland's largest university. His laboratory, which also belongs to ETH, the Swiss Federal Institute of Technology, occupies a floor and a half in a research building on the university's new campus, perched in the hills above the city center. The northern wall of Schwab's office, part of the building's all-glass northern rear, offers a breathtaking panoramic view that extends all the way to the Rhine valley and the wavy skyline of the Black Forest. Schwab, a tall, lanky man, recognized from afar at scientific conferences by his trademark silk cravat tucked into a white shirt (an American neuroscientist who copied the outfit was nicknamed "Pseudo-Martin"), had studied zoology in Basel, followed by a fellowship at Harvard Medical School, and then conducted neurobiology research at Max Planck Institute in Munich before settling in Zurich, where he now codirects his university's Brain Research Institute.

Prompted by Albert Aguayo's studies, Schwab had started looking for ways to create a growth-stimulating environment for the spinal cord in the early 1980s. The conceptual basis for his work was Cajal's idea that regeneration failed to occur in the adult central nervous system because of lack of growth-promoting substances. In fact, the first growth-stimulating molecule discovered, the nerve growth factor, was originally thought to play a role only in development, not in adult organisms. Schwab, who in his early career studied the development of nerves, had much experience with the nerve growth factor and he tried using this molecule in regeneration. But then a new growth factor was discovered in a laboratory next door to Schwab's, and this molecule came from adult brains, those of pigs. Soon after that, different growth factors were found in different areas of adult rat brains.

Schwab thought this strange because the discoveries didn't agree with Cajal's theory that the adult brain lacked growth-promoting substances. To clarify the situation, he designed what he calls his "breakthrough" experiment: He found that young neurons, which grew happily in peripheral nerve tissue, failed to grow in central nervous tissue with or without the nerve growth factor. Apparently, growth factors alone failed to solve the "enigma of paralysis." Something else was going on. Schwab realized he had found the opposite of what he was looking for: The failure to grow, he proposed, resulted from a growth-inhibiting mechanism. "I was looking for stimulators and found inhibitors," he said.

In 1985 Schwab made the risky decision to put his entire new laboratory at the University of Zurich, which then consisted of about six people, on the inhibition project. Within three years, his gamble paid off. He discovered that the inhibiting substance was made by oligodendrocytes, the octopus-shaped cells that produce myelin, the insulating sheath of nerves. "The neurons really hate oligodendrocytes, they stop if they hit upon them and grow around them. It's very dramatic and you can see it easily," Schwab says. In the next step the scientists established that the inhibiting substance was one of the proteins in myelin.

Clearly this seems paradoxical because it was already proven that myelin plays a vital role in spinal cord repair: Nerve fibers must be properly myelinated in order to conduct impulses at normal speed. However, the nerves must be coated with myelin *after* they have grown, not before; in fact, the myelin sheath probably serves as one of the barriers to further growth. From the point of view of the healthy adult nervous system, imposing such a barrier makes sense: Once the complicated machinery of the brain and spinal cord is put together, any further growth can cause havoc. Like a photograph placed in a fixative solution after it has been developed, the central nervous system, once its development is completed, keeps its fibers firmly in place. Myelin is probably one of the system's "fixatives." Support for this concept is provided by studies in different laboratories showing that nerve injuries are repaired in chick and rat embryos and in the newborn opossum, but only if these injuries are created at an early stage, before the developing organisms begin to form myelin; once the myelin sheath appears, regeneration of the injured nerves no longer takes place. The problem is that when the spinal cord is damaged, the same growth-inhibiting system becomes a barrier to regeneration. (In peripheral nerves, myelin does not interfere with regeneration. Peripheral myelin is made by different cells, the Schwann cells, and its biochemical composition is different from that of myelin in the central nervous system. Moreover, in peripheral nerves, but not in the spinal cord, myelin debris from nerve damage is removed from the injury site by scavenger immune cells.)

In the first three years of the project, Schwab's laboratory identified several basic features of the inhibiting protein, including its molecular weight, and scored a real coup. The scientists managed to produce an antibody, called IN-1, that blocked the protein's activity. Antibodies are Y-shaped proteins manufactured by the immune system. They neutralize bacteria, viruses, and other unwelcome invaders, often by latching onto the active site on the invader's surface. But scientists can make antibodies for various purposes, to detect a particular substance or block an unwanted biological process. An antibody can be used as a plug that clogs up the active sites of cells or molecules, rendering them idle, which was precisely what IN-1 did.

It was rather remarkable that Schwab's team was able to create an antibody that blocked the inhibiting protein while the protein itself remained unknown. The scientists injected pieces of rat myelin into a mouse, prompting the mouse's immune system to produce numerous antibodies against the foreign substance. Then, relying on intuition, luck, and hard work, they systematically screened these antibodies, checking which one neutralized the inhibiting effect of myelin, until they zeroed in on IN-1. Thus, Schwab's team could launch experiments with the antibody early on, years before the inhibiting protein would be identified.

In the laboratory dish, the IN-1 antibody, apparently holding the unknown inhibitor protein at bay, produced a powerful "uninhibiting" effect on nerve cells. Neurons that refused to grow in the presence of myelin-making oligodendrocytes sent off long, robust fibers when IN-1 was added to the mix. If the same effect were re-created in live animals or in the human body, IN-1 could make the injured spinal cord regrow. Although Schwab says his primary goal has been basic research, not curing paralysis, his antibody would become a beacon of hope for spinally injured people worldwide.

## Crossing Forbidden Territory

Application of the IN-1 antibody to laboratory animals was entrusted to Dr. Lisa Schnell, one of the most skillful researchers in the lab ("If something is tough, Lisa does it," says Schwab). Her career had been delayed by the relatively low status of Swiss women (who did not even gain the right to vote until the last decades of the twentieth century). Schnell's parents didn't think that, as a girl and the youngest of their six children, she needed a university education. She was trained as a medical technician, but in her early 40s, after her children had grown, she quit a steady, well-paid hospital job and realized her life-long dream of becoming a scientist when she accepted an entry-level position in Schwab's lab. The move suited her personality. With her girlish dark-red bangs and a preference for blue jeans, she felt more at ease in the informal atmosphere of a university lab than in the hierarchical "Herr Doktor" structure of a Swiss hospital. As if making up for lost years, Schnell worked on the regeneration experiment through weekends and holidays. It was around Christmas of 1988 that she made a dramatic observation in support of Schwab's inhibition theory.

Winter suits Zurich, particularly when a mantle of snow blends the city's medieval towers with the more modern structures, and when, at Christmas, thousands of warm lights decorate its Old City and the main streets. But the university laboratory where Schnell stayed late alone one night in the microscope room was untouched by the festive decor, and because of the holiday season, it was unusually quiet. The result of Schnell's day in the laboratory was a batch of several dozen glass plates covered with pale-green, zebralike stripes. Each stripe was a slice of a rat spinal cord, about 2 cm long, 3 mm wide, and a fraction of a millimeter thick, permanently pasted onto the glass.

The stripes were the culmination of an experiment that had lasted several weeks: First, Schnell and Schwab had made sure that the spinal cords of their laboratory rats received a steady supply of the IN-1 antibody; to do so, they had implanted the rats' brains with fast-growing cells designed to secrete the IN-1 antibody into the fluid bathing the brain and spinal cord. Every cell was secreting about 1,000 antibodies per second. The scientists then cut the rats' spinal cords and waited two to three weeks. Finally, the rat cords were removed and examined under a microscope for potential effects of the IN-1 antibody on nerve fiber regrowth. That memorable evening, Schnell placed one of the plates under the microscope, adjusted the viewer, peered into the lens—and gasped. "I thought, oh my God, this is it," she recalls. "I ran away from the microscope to calm down, walked down the hall to my office, then came back to the instrument." The sight that gave her such a jolt was still there.

Magnified 200 times, a humble piece of a rat spinal cord is transformed beyond belief. It fills up the entire field of vision and lights up in magnificent colors, which vary depending on the method of analysis. In Schnell's experiment, the fluorescent-orange nerve fibers stood out against a pitch-black background of the rest of the tissue, like a tail of a gigantic comet sprayed over the night sky. Because the rat's spinal cord had been severed, the comet's tail was interrupted with a gaping black hole. In reality, the hole was only some 2 millimeters wide, but under the microscope it looked like an enormous gap, its far ends light years apart. The gap broke down the communication between the brain and spinal cord, leading to paralysis.

What Schnell found so overwhelming was that she could see a few fluorescent specks, a few sparks tossed by the comet's tail, below the gap. The fluorescence is created by a tracing material that is injected into the brain and travels down the spinal cord fibers until it is stopped by the gap. If Schnell could see the tracer below the gap, this meant that some of the nerve fibers had crossed it and were crawling further down the spinal cord. She had caught these fibers in the act of regenerating.

It was a great conceptual leap for spinal cord regeneration. In earlier experiments, such as those conducted by Albert Aguayo's team in the 1980s, fibers had regenerated only through growth-friendly "sleeves" of peripheral nerve; they stopped when they reached the end of the bridge and refused to enter the other stump of the spinal cord. Now, for the first time, the fibers had regrown in forbidden territory, the central nervous tissue itself—apparently, because the antibody neutralized the inhibitor in myelin.

The following morning Schnell called the only two colleagues who also worked during the holidays to look at her results. They closed the door to the small microscope room so as to see the fibers better in the dark. One of them peered into the microscope and said, "You call this regeneration?" He was referring to the small number of growing fibers, only about 5 percent of those that had been cut. Schnell was also disappointed but not discouraged: "Of course, I was also hoping to see 70 percent of the fibers regenerating, but just because I expected 70 doesn't mean I won't take 5."

Schnell's dedication to the project took unconventional forms. Processing the brain and spinal cord tissue was a lengthy and tricky business and on some occasions, when an evening outing to a concert interrupted the experiment at a crucial stage, Schnell packed the vials with rat tissue into her bag and during the intermission washed the samples in the theater's ladies' room. If the procedure attracted bewildered looks from proper Swiss ladies using the facilities, she tried to act casual, as if washing pieces of animal tissue during a concert intermission was a perfectly normal thing to do: "I pretended, like, don't you do that too, when do you take care of *your* spinal cords?" To speed another experiment, Schnell took rat tissue samples home. The tissue was bathed in a sucrose solution whose concentration needed to be gradually increased, and Schnell replaced the solution several times during the night. Her two children got used to finding vials with rat brains and spinal cords in the kitchen refrigerator next to their Swiss cheese. They relished the opportunity to shock friends with the weird stuff, but most important, the practice allowed Schnell to carry on with the experiment literally around the clock.

In the year that followed Schnell's remarkable observation in the microscope room, she and Schwab repeated the experiment five times, comparing animals that had been treated with the IN-1 antibody and those that were treated with an antibody against a different protein, the so-called controls. The animals were coded so that the scientists didn't know which ones they were looking at. "You are sweating at the microscope, counting the fibers, and you tend to formulate a hypothesis that this is probably a control, and you have to continuously fight this in order to be objective. In the end, you break the code to see which animals were treated. If the result fits, it's a great moment, and if it doesn't it's really bad," Schwab says. In that study, every time they broke the code it showed that the IN-1 antibody facilitated regeneration. Severed fibers in rats treated with the IN-1 antibody grew 11 millimeters, compared with only 1 millimeter in animals belonging to the control group. "Initially, I thought this project was very high-risk and we were astonished that it worked," Schwab says.

The results, reported in *Nature* in January of 1990, were widely covered by the media, generating dozens of calls from spinally injured people. One of the callers was Zurich businessman Ulrich Schellenberg, an advertising executive who had been paralyzed in an accident 12 years earlier and wanted to channel his creative energies into stimulating paralysis research. After reading about Schwab's results in the *New York Times*, Schellenberg, in collaboration with a prominent Swiss physician, went on to create the International Research Institute for Paraplegia, today one of Europe's largest private organizations to fund spinal regeneration research.

## An Antibody for Humans

Schwab's antibody research was certainly not the first study to inspire hopes of a potential cure for paraplegia, but it offered the promise of a regeneration therapy based on solid scientific ground. Schwab was therefore a natural choice to become the first recipient of the award named after the *Superman* star Christopher Reeve.

Professor Schwab has received many prestigious honors, including the Wakeman Award, but getting the Christopher Reeve Research Medal was the most moving experience of all. The medal was created in conjunction with the establishment of the Reeve-Irvine Research Center for studies into spinal cord injuries at the University of California, Irvine. It was presented to Schwab in September 1996 by Reeve and his wife Dana at an international equestrian competition in San Juan Capistrano, California.

The event was conducted in a lavish Hollywood style—"all very American," according to Schwab—with a "dinner under the stars" in the courtyard of a Spanish monastery, attended by real-life movie stars. But the ones to steal the show were the scientists. Schwab had asked to give a presentation about his team's research and had brought with him slides with tissue samples, which had to be projected onto a screen for the benefit of Reeve, whose injury prevents him from peering into a microscope. Reeve had just arrived in a private jet and was so tired that his assistant warned Schwab he would not stay long. However, Reeve ended up attending the presentation for three hours, getting from Schwab a first-hand account of scientific progress and of the kind of hope he could realistically have.

At the time, the IN-1 antibody was not ready for human use, and in fact, scientists are still working on this goal. First, they must produce a "humanized" version of the animal antibody and create smaller, more effective versions of the molecule. The antibody must be tested for safety, to make sure it will cause no chronic pain or spastic movements by facilitating the growth of the wrong nerve fibers. "The majority of people with paralysis are very young. They

may live for 30, 50 years. These are not terminally ill cancer patients, so if you create a pain problem, this would be a total catas-
trophe," says Schwab. "Therefore, the safety requirements for any
treatment have to be extremely high, and this cannot be done in a
few months." Also, a safe system must be developed to continuously deliver the antibody to the injury site. (The method used in
the rat study, the release of the antibody via proliferating cells implanted into the rats' brains, is obviously unacceptable for humans.)

Then there is the question of efficacy. In November 1995, Barbara Bregman of Georgetown University, a veteran and respected
regeneration researcher, reported in *Nature* on a joint study with
Schwab's laboratory, in which the antibody helped paralyzed rats
recover partial use of their hind legs. This was the first time that
recovery from paralysis was observed in rigorously conducted research, and the accompanying editorial said that "cautious optimism is now possible." However, the scientists had not cut the rats'
spinal cords completely, in order to leave a surface for the cut fibers
to grow upon. The researchers were well aware that rats can regain
movement with a very small number of spared fibers and took great
care to check whether recovery was indeed due to regeneration, yet
the *Nature* editorial cautioned that other mechanisms, such as compensatory sprouting of uninjured neurons, might have contributed
to recovery. Will the antibody work in humans, whose central nervous system is substantially different from that of rats? Schwab believes that final efficacy and safety tests must come from trials on
monkeys, which are closer to humans than are rats. Such experiments were launched in Switzerland in the year 2000. (In these trials, scientists damage one small region of the monkey spinal cord,
which controls fine movements of the hand. While the animals receive the antibody treatment, they can move about freely and their
function is not severely impaired.)

Schwab also warns that any recovery with his antibody is likely
to be partial because the number of regenerating fibers is small.
This could be due to the presence of other growth inhibitors in
spinal cord tissue that are not neutralized by IN-1. It is also possible that not all neurons are capable of growing, even under the

best conditions, or that the antibody needs to be improved further. In 2000, Schwab began to develop an antibody for humans with a new commercial partner, the pharmaceutical giant Novartis. Meanwhile, another team in his laboratory was looking for the inhibiting protein that the antibody was supposed to block. The hunt for the protein turned out to be trickier than anyone had imagined, and the scientists were wondering if they were ever going to find the elusive molecule.

# 8

# Overcoming an
# Invisible Barrier

T he way scientists look for an unknown protein can be likened to the strategy for capturing an imaginary lion in the desert. The strategy, invented by mathematicians as a humorous metaphor for certain problem-solving approaches, consists of dividing the desert in half, checking which half contains the lion and dividing it again in two, then repeating the process until the area available to the lion is so small he has to balance precariously on one paw. At this stage, all the lion hunter needs to do is place a cage next to the lion and gently push the king of beasts until he tips over into the cage. Similarly, biologists divide their source material into parts, check each part, and break it down further until they capture—in technical language, "purify"—a single substance that does what their protein is supposed to do.

When Martin Schwab first proposed that the spinal cord contained a growth-inhibiting molecule, the inhibitor appeared to hide alluringly close. Yet whenever scientists tried to physically trap the molecule using the "lion" strategy, it kept fleeing. Tracking it down turned into a detective story with a few heart-stopping twists.

The search had begun in 1988, and the person in charge was Christine Bandtlow, a young, sharp, German neuroscientist who had just joined Schwab's laboratory after completing her Ph.D. studies on brain growth factors at Max Planck Institute in Munich. She took the risk of joining a pioneering project because she was excited about its novelty and, having gotten to know Schwab at Max Planck, trusted his ability to launch a serious investigation. In her candid manner, Bandtlow, a vivacious, witty woman, recalls the ups and downs of the protein hunt. "I thought, well, with this fantastic antibody and the knowledge of the protein's approximate molecular weight, it would take us two, three years, not longer," she says. The beginning indeed seemed to augur well. She quickly determined that the inhibitor was not among the five or six major proteins that make up 90 percent of myelin, which narrowed the search to the remaining 10 percent. But this 10 percent still contained hundreds of proteins, and looking for a scarce molecule soon revealed itself as a major handicap. "You try to separate your protein from the rest, and each time you do a separation you lose a lot of material, and all along the way you don't know what you're looking for. You divide the material by 10, and you still have 500 proteins. It's really awful," says Schwab.

After working for more than three years with the help of two technicians, Bandtlow still had nothing. That meant no scientific publications, a potential death sentence to a young researcher's career in the "publish or perish" world of science. The elusive molecule received the nickname "The Career Inhibitor." Bandtlow began to pursue what she called bread-and-butter projects, parallel lines of research that seemed less innovative but more secure in terms of results. At the same time, the team began to feel the heat of competition. "At the beginning, we may have been lucky because we scared off competitors. Everybody thought we were so advanced that it was a piece of cake for us to finish," Bandtlow says.

However, over the years other scientists started looking for inhibitory proteins.

In the fall of 1994, the journal *Neuron* dropped a bombshell. It published papers from two independent research teams, both announcing that a protein called MAG, found in myelin, blocked the growth of neurons in the central nervous system. "They had a title that we thought we would write for *our* protein—'Such and such a molecule is a major inhibitor of regeneration,'" Bandtlow says. One of the teams had not even been looking for an inhibitor; they stumbled upon it by chance while studying the synthesis of myelin.

"That was an extreme shock for us at the time. We thought our project was finished, but ultimately the MAG story turned out to be extremely useful," Bandtlow says. The *Neuron* reports indicated that growth inhibition in the spinal cord was a complicated, multi-factor mechanism. Moreover, they helped reveal a fascinating property of nerve cells: their responsiveness changes with age, so that the same protein can have two opposite effects at different stages in a neuron's life. The Swiss team happened to be working with young neurons, in which MAG had no effect, while the two other teams used older neurons that are easier to inhibit, which enabled the scientists to detect the inhibition produced by MAG. These revelations would later channel Bandtlow's career in a new direction. Now a professor of neuroscience at the University of Innsbruck in Austria, she studies the signaling mechanisms that control the response of neurons to growth-promoting or -inhibiting factors.

Further studies confirmed that although MAG was indeed an important inhibitor of growth, it was certainly not the only one. The Swiss scientists had evidence that *their* inhibitor was still at large: MAG did not interact with the IN-1 antibody, which produced an "uninhibiting" effect on myelin; this meant that the antibody interacted with another, still unidentified protein.

## Of Proteins and Alpine Peaks

In the hope that the sheer bulk of source tissue would speed up the search, Bandtlow decided to switch from rat to cow spinal cord.

She made arrangements with a slaughterhouse in Zurich, and since the best results were obtained when the tissue was fresh, the research schedule was adjusted to the schedule of the slaughterhouse.

At about the same time, the team got a new member who possessed two character traits important for a scientist: independent thinking and a high frustration threshold. Adrian Spillmann, who joined the team as a Ph.D. student in the fall of 1993, has the quiet but determined air of a person not easily swayed from his course. The slim, blond young man had considered studying economics but opted for a Ph.D. in neurobiology when he got an opportunity to work on Schwab's project. Spillmann was told, in the optimistic spirit of the lab, that the purification of the protein would be a matter of a few months. But two years later, with the search in its eighth year, there was still no protein in sight. The protein was so scarce that Spillmann would start out with more than a pound (half a kilo) of cow cord, but by the time he went through all the purification steps he ended up with barely 1 microgram of material, 20 times less than necessary to study its properties properly. "He'd just go through euphoric periods when he thought he had it, and then he couldn't reproduce the damn results," Schwab says.

Schwab jokes that Spillmann eventually snagged the protein because of his Swiss Alps origins. "People in that region are tough and not easily frustrated, they rarely give up," he says. Spillmann, who had spent his early childhood in a small alpine hamlet, is not sure this biographical detail was relevant to the project, but he concedes that the mountains do shape one's character to some degree: Their sheer height may be setting a scale that dwarfs other obstacles. "It's the mountains standing in front of you, going up thousands of meters, and there's no way you can move them," he says. The mysterious protein became Spillmann's mountain, his very own alpine peak.

Because of the risk that the scarce protein could be destroyed by heat-activated enzymes, much of the work was conducted at 40°F (4°C). For weeks on end, Spillmann would don a sweater or a ski jacket even during the summer and spend four to five hours a day in a chilly, windowless room. He tried every purification method in the book, sometimes resorting to trial and error. "To be honest," Spillmann recalls, "it would be a lie to say I was never frustrated, but I felt there was something in this sauce that had a strong inhibitory effect." He would occasionally drive his supervisors up the wall by listening to them, then doing the experiments his own way. "I knew I made myself look a bit stubborn, but my decision was, don't listen to other people, go your own way, because, if they knew what was correct, they would have found the protein already," he says. Among the factors that kept Spillmann going were Schwab's contagious enthusiasm and the support of his fellow team members, who were always willing to take him out for a beer when the project seemed to be going nowhere. He needed that beer about once a week, for nearly four years.

Toward the end of 1997, Spillmann finally purified the protein, but the laboratory was not in the mood for celebrations. The scientists still had doubts whether they had purified a major regeneration inhibitor. They hurried to clone the protein's gene, the piece of DNA that contains the blueprint for the protein. Knowing the gene makes it possible to manufacture the protein in large amounts for extensive experimentation. The cloning had to be done fast because once the protein was described in a scientific report, any other laboratory could embark on the cloning and beat the Swiss scientists to the task. In fact, two other teams, one at Yale University and another at a pharmaceutical company, would soon enter the race to identify the inhibitor's gene. Spillmann performed part of the cloning, an intellectually rewarding task that had originally attracted him to the project, but after finishing his Ph.D., he went into economics, which appealed to him as a more secure way of attaining professional fulfillment than science. He says his decision to quit was unrelated to the difficulty of the inhibition project. "If I had chosen a less demanding Ph.D. I might have quit earlier," he says. "I stayed on because the search for the protein was challenging, and it was going to be a step toward a cure for paralysis. At the time, the thinking still was that this one protein was switching off the regeneration in the whole system, and I was proud to be part of such an important project."

As the cloning of the gene proceeded, the possibility that the laboratory had purified the wrong molecule began to loom large: The gene did not look right. Most disturbingly, it did not seem to contain the code for a surface molecule, yet a growth-inhibiting protein would be expected to sit on the surface of cells. "Horror reigned in the lab that we had purified the wrong thing," Schwab says. "I can laugh about it now but then it was pretty terrible." While putting the protein through a battery of tests for two more years, the scientists found no explanation for the molecule's bizarre properties, but they did confirm that it blocked regeneration.

The final, festive step was to give the new protein a name, and Schwab promised two bottles of champagne for the best candidate. Scientists approach the name-giving of molecules with as much trepidation as parents choosing a name for their new offspring. They look for a catchy moniker that will contribute to their molecule's popularity and, if possible, reflect its activity. For example, an abnormality in the protein called Dissatisfaction disrupts the mating behavior of fruit flies, leaving both male and female flies unhappy; Yotiaio, a scaffold protein of nerve cells, is named after a Chinese breakfast noodle on account of its long and stringy structure.

Schwab never had to buy the champagne because he came up with the name himself. In the scientific report that appeared in *Nature* on January 27, 2000, he called the inhibitory protein Nogo, echoing the commands "go" and "no go" used in primate behavior tests. (In scientific convention, the name of the gene is usually written in italics: *nogo*. The protein made by the gene bears the same name but, to distinguish it from the gene, is spelled in roman font and sometimes with a capital letter: Nogo.) Two other contenders in the race to clone Nogo, the Yale scientists and the commercial team, reached the finish line at the same time with Schwab and published their findings in the same issue of *Nature*.

The cloning of the *nogo* gene has opened exciting new possibilities. It makes it possible to produce the protein in large quantities and facilitates the development of better antibodies for human use. And in January of 2001, Yale University researchers reported that they had identified one of Nogo's receptors, a sticky protruding molecule on the surface of nerve cells that allows them to make contact with the Nogo protein. This finding may help scientists devise alternative regeneration therapies that would block the receptor, making the cells invisible to Nogo. However, in the 15 years that it took to nail down Nogo, the regeneration barrier turned out to be infinitely more complicated than originally thought.

## A Tank of Sharks

Gone are the days when scientists could hope that one molecule would unlock the secret of regeneration. Rather, the barrier to growth is more like a tank teaming with sharks. Myelin clearly contains at least two inhibitors, Nogo and MAG, and possibly many others. Moreover, outside of myelin, central nervous tissue may be strewn with leftover guidance molecules that were used to wire up the nervous system in the fetus; in adulthood, these molecules may turn into inhibitors of fiber growth.

Further complicating the picture, an old enemy of regeneration, the scar, has made a comeback in a new guise. In the 1950s, William Windle believed that the nervous system's supporting cells, the glia, physically blocked the growth of regenerating nerve fibers by creating a scar. In the modern version of this theory, the glial scar is also a molecular barrier. Glial cells, particularly the star-shaped astrocytes, proliferate at the site of the injury in preparation for creating the physical scar, but before they form a mechanical barrier, they release a host of chemicals, most notably large protein-sugar molecules called proteoglycans. These molecules float in the goo called the extracellular matrix, which holds cells together, and contribute to the hostile environment that blocks the attempts of nerve fibers to regenerate.

After Windle's time the scar idea fell out of favor, and in the 1990s its advocates had difficulty bringing it back. One of the scar's main champions has been Jerry Silver of Case Western Reserve University in Cleveland, Ohio. Silver, who apart from a fellowship at Harvard Medical School has always worked at Case Western, his

Ph.D. alma mater, already has one scar victory to his credit. He had originally studied the development of the fetal nervous system and in the late 1980s developed a strategy for bridging injured rat spinal cords with fetal astrocytes. Silver cofounded a company, Gliatech; in one of the company's spin-off projects he, together with Gliatech scientists, developed a gel against glial scars that sometimes form around peripheral nerves after surgery, causing pain and restricting movement. The gel was approved by the U.S. Food and Drug Administration for preventing scarring in certain types of low back surgery, but Silver parted ways with Gliatech and returned full-time to university research. He continued to work on astrocytes and focused on the role they play in chemical and physical scarring after spinal cord injury. By the mid-1990s he reached the conclusion that the scar was the major culprit in the lack of spinal cord regeneration. Silver, an enthusiastic and easily excitable man, speaks of science as a battlefield where new ideas must be backed by heavy artillery. He regarded the scar as an embattled concept that would never get recognized without a good fight. "If you go against a giant you need big guns," he said, referring to Martin Schwab, whose myelin inhibition concept reigned supreme in the regeneration field.

By the late 1990s, Dr. Silver had gathered the scientific ammunition in support of the chemical scar in the injured spinal cord. Together with colleagues he published two major studies designed to distinguish between the effects of the scar and those of myelin. In one study, nerve cells were inserted into the spinal cord so gently that the transplantation caused no scarring, and these cells sprouted fibers that grew at an impressive rate of 1 millimeter per day despite the presence of myelin. In contrast, when the fibers encountered a cut in the spinal cord, they stopped growing and their tips withered, a hallmark of regeneration failure demonstrated by Cajal 100 years earlier. Silver found that the area of the cut, particularly around the malformed fiber tips, was full of proteoglycans, the molecules believed to form the scar.

The interpretation of these results has been a subject of bitter controversy that occasionally provokes heated exchanges at scientific conferences. Dr. Silver believes his studies show that myelin is not the predominant obstacle to spinal cord regeneration. Myelin advocates strike back by arguing that in Silver's experiments the neurons were transplanted so gently that they left myelin undisturbed, so that myelin's inhibiting proteins were not exposed. In fact, the functions of the two barriers are by no means mutually exclusive. On the contrary, both barriers probably need to be overcome for regeneration to take place, and scientists are designing strategies to eliminate them both.

The chemical scar, owing to the work of Silver and of other scientists, is now recognized as a potentially important roadblock to regeneration, and it is studied by a growing number of laboratories, but in 2001 the scar's exact makeup was still uncertain and scientists had only begun to develop approaches for effectively fighting it. In the meantime, at least three strategies had already been designed to overcome all possible myelin inhibitors in one sweep. Canadian researchers headed by Dr. John Steeves, at the University of British Columbia in Vancouver, had developed a treatment that temporarily strips nerve fibers of myelin, allowing them to regenerate, then makes sure that the protective myelin sheath is restored. Potential future applications of the treatment in humans are being explored by a Vancouver biotech company, Neuro Therapeutics Inc. Another Canadian team, headed by Dr. Lisa McKerracher of the University of Montreal and Dr. Samuel David of McGill University, developed an approach that neutralizes myelin as a whole. The scientists call it a therapeutic vaccine because it works much like vaccination against infectious diseases: The immune system is prompted to produce antibodies, only here, instead of infectious organisms, the antibodies are aimed at myelin. In laboratory mice, the vaccine has produced a massive regrowth of cut fibers after spinal cord injury. And a newer approach is to look for molecular "switches" that regulate inhibition. By 2001, at least three teams—led by Dr. McKerracher at University of Montreal, by Dr. Marie Filbin at Hunter College, and by Dr. Larry Benowitz at Harvard University—had discovered such switches and were trying to figure out how to turn them off in order to inhibit the inhibitors.

Schwab's IN-1 antibody approach is more limited but, as the oldest therapy, it is likely to finish first. Like most other methods, it promotes the growth of 5 to 10 percent of fibers, but even this small number can make a difference in the function of a person with paralysis. "We must respect the human body. It has a tremendous ability to recover," says the prominent American neuroscientist Dennis Choi. "A little push, a little boost, like Martin Schwab's approach, might just be enough."

This new confidence is bolstered by a growing number of experiments proving that, at least in laboratory rats, recovery of function after paralysis is possible. One of the most talked about studies was conducted some 900 miles north of Zurich, at the Karolinska Institute in Sweden.

# 9

# The Rats Are Walking

The Karolinska Institute occupies a hilly landscaped campus on the northern outskirts of Stockholm. In contrast to the picturesque city center, the Institute's architecture is rather grim, a collection of boxy, dark-red brick buildings lined with endless rows of small, white-framed windows. Inside, however, the laboratories and offices are airy and spacious, their interior pleasing the eye with the clean lines and colors characteristic of Scandinavian design. Now the Karolinska is known as a foremost center of medical research and training and as the institution that awards the Nobel Prize in physiology or medicine, but when it was established by royal decree some 200 years ago its goal was "the enlightenment of skilled army surgeons." In many countries surgery was then a craft practiced by barbers who, the Swedish king felt, needed to be "enlightened" by the study of medicine. It is historically fitting that the most remarkable surgical repair of the animal spinal cord should have taken place at the Karolinska.

On July 26, 1996, a team headed by Lars Olson produced the first convincing evidence of recovery in rats whose spinal cord had been completely cut. The rats had undergone complicated surgery performed in Olson's laboratory by the Taiwanese doctor-researcher Henrich Cheng. The editorial that accompanied the scientific report in *Science* hailed the study as a "milestone" and announced that effective spinal regeneration for humans was "no longer a speculation but a realistic goal." The popular media gave major play to the story, probably because it followed by only a year the injury of the *Superman* actor Christopher Reeve, which had stirred tremendous public interest in spinal cord research. The Karolinska rats made front-page news around the world and were featured on television network news. "It's doubtful that would have happened before Reeve brought the cause to national and international attention," *Time* magazine wrote. "After the Stockholm success, [Reeve] smiled and looked eager. 'If that's what they are doing over there,' he said, bring me to them. I'm a rat.'"

Olson was concerned about the media blitz. "There's been too much hype about our results," he says. "I'm not ashamed to give spinally injured people some hope, on the contrary, but I don't want to generate false expectations of a rapid cure." Yet the study created what has been described as "a throb of excitement" in people with paralysis worldwide. Finally, a proof that regeneration can work! At a workshop on spinal cord injury organized by the U.S. National Institutes of Health in the fall of 1996, Kent Waldrep, head of a private foundation that supports spinal cord research, presented workshop organizers with T-shirts showing a cartoon rat prancing on its hind legs, with the caption, "The Rats Are Walking!"—a reminder of the expectations ignited by the animal research.

Professor Olson did not need to be reminded about the meaning of his work. Some overseas colleagues describe him as "very Swedish," referring to his icy Nordic manner, yet he has been uncharacteristically forthcoming on the emotional reason that brought him to spinal cord research. When he started studying medicine at the Karolinska in the early 1960s, he was initially inspired by his charismatic teacher, the eminent neurobiologist Nils-Åke Hillarp, to specialize in research of Parkinson's disease, an area in which Swedish neuroscience has a strong tradition. But then a friend of Olson's from medical school dived into a river headfirst to celebrate the passing of the anatomy examination. The river was shallow, the young man broke his neck, and has remained paralyzed. Since then, Olson has developed an interest in the spinal cord.

Because he has always worked at the Karolinska and plans to retire one day from the same department he once joined as a student, Olson describes his professional life as "monotonous," yet his scientific record has been anything but. In the early 1970s, he performed some of the first modern-day transplants of brain tissue in laboratory animals. Among his most impressive experiments was the grafting of chunks of embryonic brain into a rat's eye, where the development of these brain segments could be observed as if under a transparent dome. But in the back of Olson's mind was the thought that one day he wanted to apply this research to human disease. "That's why we have our jobs," he says.

To people outside the scientific community the idea of brain tissue transplants seemed at the time a bit too daring, if not plain weird. It evoked the eerie prospect of a whole-brain graft, or as a 1980s *Newsweek* article put it, "the ultimate transplant." However, at least one brain disorder lent itself perfectly to treatment with grafted tissue, and it became the test case for the innovative approach. Parkinson's disease is caused by a loss of cells in one tiny region of the brain, the substantia nigra. These cells make dopamine, the brain chemical responsible for movement control. Scientists hypothesized that a graft could compensate for the loss by supplying dopamine-making cells from the adrenal glands or fetal tissue (or, as later discoveries would suggest, by immature cells known as stem cells).

Originally, in the early 1980s, several people with Parkinson's disease had small pieces of their own adrenal glands transplanted into their brains, but ultimately fetal tissue proved more promising. The first human trial of fetal transplants in Parkinson's disease was a fine example of a common goal transcending individual ambitions. In the 1970s two teams that successfully accomplished fetal transplants in animals—one led by Olson and another by researchers at the University of Lund, in the south of Sweden—had competed with each other. But when the Lund scientists got a local hospital to agree to the controversial procedure in humans in 1987, competition was cast aside; Olson hired a private jet to bring additional fetal tissue and his team from the Karolinska to Lund to participate in the clinical trial. (A human transplant required a substantial amount of fresh fetal tissue that could only be obtained from several fetuses, yet ethical guidelines did not allow the timing of abortions to be dictated by the transplant program.)

In some people, the efforts paid off. "It's a fantastic thing to see a person with Parkinson's, who is lying stiff in bed, walk up to you four or five months after grafting. This is the best thing that can happen to you as a scientist, when your animal research finally leads to something good for a human being," Olson says. Fetal transplants have since been performed on several hundred people with Parkinson's worldwide and are believed to alleviate the symptoms of the disease in some cases, although the overall safety and efficacy of the procedure still need to be improved. Olson, in the meantime, turned to basic research in other areas in order to "stay at the absolute frontier and invent new things." One of these areas was spinal cord regeneration.

Because of the respect we have for the brain as our thinking organ, we tend to view brain disorders as the pinnacle of complexity, but the damaged spinal cord can present an even greater challenge. The treatment of Parkinson's disease requires correcting a malfunction in one chemical system, but in contrast, repairing the spinal cord involves restoring a vast number of complicated nerve circuits that ensure communication between different body organs. "I don't want to trivialize transplantation in Parkinson's disease, but spinal cord injuries are 100 times more difficult—or actually, 1,000 times more difficult," Olson says.

# A Dumpling Lesson

Olson's research on spinal cord regeneration took an unusual turn in the early 1990s when Henrich Cheng, a virtuoso neurosurgeon from the Veterans General Hospital in Taipei, Taiwan, came to his laboratory for his obligatory year of foreign training. Olson taught Dr. Cheng the molecular basics of regeneration, his main topic of interest, and together they set about developing repair strategies that made optimal use of Cheng's surgical skills.

Dr. Cheng, a tall, stocky, bespectacled man in his early 40s, is an accomplished storyteller, who willingly shares with the listener the most amusing episodes of his career. Cheng had been a prodigy child painter preparing to become a professional artist, but he gave up on his childhood dream when his father insisted on sending him to medical school. He resented the choice until he came upon a biography of Harvey Cushing, the father of American neurosurgery, who became his hero. Cushing was also an artist, and Cheng copied him in every way he could, particularly in the care with which, according to the biography, Cushing studied every nerve during his training in anatomy. For six months, Cheng did anatomical dissections day and night, often till 4 o'clock in the morning, while paying special attention to the nervous system. During his residency, after witnessing the fate of two people paralyzed from the neck down, he developed a special interest in the spinal cord. His painting experience, however, had not been wasted. In fact, if Cheng had his way, he would send all surgeons to study art. The classical school of painting, he says, helps develop a feeling of space because it trains the eye to study objects in three dimensions and the interplay between distance, color, shadow, and light. "I treat my surgery as an art, and I find the same joy in finishing an operation as I do in finishing a painting," he adds.

One of Cheng's first ambitious projects at the Karolinska was the shortening of the vertebral column, the bony pillar that encases the spinal cord. In mammals, the spinal cord comes under tension during development: Bone grows faster than do nerve fibers; as a result, the adult spinal cord ends up being shorter than the full length of the adult vertebral column. Also, the cord is somewhat elastic, possibly to allow for greater flexibility when the vertebral column is bent. Consequently, when the cord is cut, its two ends snap apart like pieces of a torn rubber band, creating a gap. To make the ends touch again, Cheng removed half of one vertebra and the opposing half of the adjacent one. Working with the team's engineer, he created a metal device with screws, nuts, and bolts that slowly brought the two halves of the vertebrae together over the course of two or three days. During this time the postoperative swelling in the spinal cord stumps subsided and they neatly came into contact without being pressed together.

The major outcome of this technical feat was that Cheng learned Swedish, having spent long hours with the engineer, who spoke neither English nor Chinese. Although some fibers did regenerate in the rat spinal cords fused with the help of the metal device, there were not enough to produce recovery of function. The scientists then thought of using peripheral nerve bridges, the same technique as the one used by Albert Aguayo and his colleagues in Montreal 15 years earlier. When Aguayo's team had reported its results in the early 1980s, convincing the scientific community that cut spinal cord fibers could regrow, Olson paid little attention. He already believed regeneration was real because he had seen it in the brain during his earlier grafting experiments in the 1970s. Now, however, Olson's laboratory would bring Aguayo's technique to its ultimate crowning, the recovery of function.

Cheng and Olson made several important additions to the bridging methods. Back in Taiwan, Cheng had tried suturing the cut spinal cords of experimental animals but found the cord was too soft and fragile to hold the stitches. Now he decided to hold his bridges in place by fibrin glue, the sticky protein substance that helps blood to clot, rather than by sutures. Another major innovation was the addition of a growth factor. Olson had been studying growth factors since the 1970s, and he believed that they play a key role in regeneration. By the early 1990s, numerous growth factors had been discovered and their role in nerve growth had been well established. But how to deliver them to the regenerating nerve fibers?

The solution was born out of nostalgia. Cheng and his wife liked Sweden, especially its clean environment and abundant space, which formed such a sharp contrast with their native Taiwan, but they missed some staple Chinese foods. In particular, in Taiwan they used to prepare dumplings using ready-made dough envelopes, which were unavailable in Sweden. Cheng's wife suggested that he invite a colleague from the less-industrialized mainland China, where prefabricated food is not commonly available, to teach them to make dumpling dough. The colleague, Dr. Yihai Cao, worked in the molecular biology department at the Karolinska where he used a chemical known to stimulate the growth of nerves and other tissues called acidic fibroblast growth factor.

As Cheng was preparing the dumplings' meat filling and talking about his nerve-grafting experiments with the fibrin glue, Cao said the dough envelopes were ready to be stuffed. At that moment Cheng had the idea of using the dumpling model in his spinal repair operation: The fibrin glue would serve as the envelope, and the growth factor would be the "meat." The following morning he rushed to Olson's office, they checked whether the glue would hold the "filling," and they went on to develop a slow-release mechanism for the growth factor. Dr. Cao was rewarded for the dumpling lesson by becoming a collaborator in the study and later a coauthor of the scientific paper that reported its results.

# Idea No. 303

What eventually helped paralyzed animals walk was yet another of Cheng's eureka! moments. Many scientists had found that the fibers regenerated beautifully through the peripheral nerve "sleeves" but stopped when they reached the end of the bridge. The growth, as Martin Schwab had shown, was apparently blocked by inhibitor molecules in the white matter of the spinal cord.

The white matter is a soft, milky substance that lies on the outside of the spinal cord. It consists of nerve fibers covered with myelin, hence its color. The inner part of the cord is made of gray matter, which largely consists of nerve cell bodies that serve as relay stations for the transmitted information. Long nerve fibers run for the most part inside the white matter, but their tips dive into the gray matter, where signals are passed to a relay station. For example, when we want to move a leg, some of the commands from the brain travel mainly down myelinated fibers in the outer part of the spinal cord, but in the area of the lower back they are relayed to motor neurons in the gray matter, which pass the signals further to the leg muscles. The gray matter contains no myelin, hence no or few inhibiting molecules.

Cheng spent two weeks thinking of a way to overcome the inhibitor molecules in the white matter. He knew that no human antibodies against myelin inhibitors were yet available and was looking for a surgical solution. One night, as he was about to fall asleep, he had an idea that caused him to jump out of bed. Since he'd been a student, Cheng had kept small notebooks in which he neatly jotted down hundreds of ideas, some related to research, others to the column on social and political matters that he wrote under a pseudonym for a major Taiwan daily. That night in Stockholm he wrote down Idea No. 303: He would reroute the bridges from the white matter directly to the gray-matter relay stations—a simple but brilliant solution that would have many scientists wondering why no one had thought of it before. (In all previous experiments the bridges had connected white matter above the injury with the white matter below.) The bridges were designed to bypass the myelin, guiding the growth of the nerve cells toward a region where regeneration is permitted.

In every experimental animal, Cheng first removed a 5-millimeter chunk of the spinal cord, creating a gap that left no doubt that no nerve fibers had been spared. He then closed the gap with 18 microscopic bridges, the maximum number he managed to fit in, all made of hair-thin peripheral nerves removed from the rats' chest muscles. To appreciate the mastery involved, picture the rat spinal cord, which is only 3 to 4 millimeters thick, about the size of a thin shoelace. "It sounds crazy, like making a sculpture out of hair, but

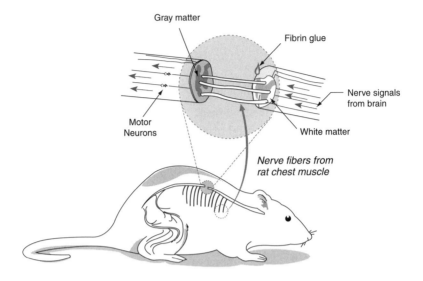

Diagram of the exquisite microscopic operation performed by Henrich Cheng at Sweden's Karolinska Institute. Cheng reconnected a rat's severed spinal cord by placing nerve fibers removed from the rat's chest into the gap. The fibers, attached to the two severed ends of the spinal cord with fibrin glue, formed a bridge between the cord's white matter and gray matter. They inspired nerve growth and acted as conduits for the regenerating axons. As a result of the operation and the addition of a growth factor, the rats regained partial use of their paralyzed hind limbs. (Adapted from an illustration by John W. Karapelou, ©1997.)

using a microscope and fine instruments it's possible," Cheng says. The difficulty was compounded by the softness of the cord and nerve tissues. "It was a real tour de force—like taking hot spaghetti, all wobbly, and trying to spoke it into the smallest, wobbly sausage," says one veteran regeneration researcher.

The bridges ran in a crisscross pattern: The ones that started in the white matter above the injury were connected to the gray matter in the opposite stump of the spinal cord—they would transmit signals from the brain to the lower body; intersecting bridges running in the opposite direction started in the white matter below the injury and terminated in the gray matter of the stump that was closer to the head—they were aimed at transmitting sensory input from the lower body to the brain. Cheng filled this amazing architectural construction with drops of fibrin glue that hardened with time, anchoring the slender bridges in place, much like liquid plaster poured over a broken limb to create a cast. In addition, he stabilized the area with metal loops hooked to the ribs to prevent a mini movement of the vertebral column from dislodging his creation.

Several weeks later, for the first time since he'd arrived in Sweden, Cheng noticed that some rats began recovering the use of their hind legs and continued to improve. The proud statement on Kent Waldrep's T-shirts was exaggerated, as they never went back to walking normally, but they could flex and move their hind legs and place their weight on them. Before the scientists announced their results, they needed to make sure this was not another case of a chance recovery, the kind that had spurred false hopes so many times. It took more than a year of testing until no doubt remained: Recovery took place only in rats treated with the crisscrossed bridges and the growth factor, but not in the controls. Tracing studies confirmed that nerve fibers in the spinal cords of the treated rats had indeed regenerated along the bridges and beyond. Seeing that the treatment worked, Olson told a reporter, "was the happiest moment of my scientific life."

## What Do Rats Think?

The study prompted scientists in different countries to try and repeat the feat, but this proved exceedingly difficult. Some gave up after a few failures, but at least two laboratories, one at the Miami Project and another at the University of Toronto, persisted for several years, despite the growing frustration with the wasted time and money. By mid-2001, no one had managed to repeat the Swedish 1996 results—apart from a former student of Dr. Cheng, the Taiwanese veterinarian Dr. Yu-Shang Lee. While working toward a Ph.D. degree at the University of California, Irvine, Lee, with his colleagues, reportedly managed to reproduce the Karolinska study and was going to present his findings at the annual meeting of the Society for Neuroscience in the fall of 2001. Was the Olson–Cheng procedure so difficult to replicate because other researchers missed some seemingly minor but significant details? Or because no one had reached Cheng's mastery in microscopic surgery? Cheng believes other scientists lacked practice. "They came to watch the technique for one afternoon, went back, and tried it, I don't think that's enough," he says. Such reliance on outstanding surgical skills is an obvious obstacle to worldwide use of a potential therapy. Still, the Swedish study provided a proof of principle and boosted the confidence of spinal regeneration researchers everywhere.

As for the study authors themselves, they went on to further explore the approach's potential. Dr. Cheng gained enough confidence from his rat experiments to try the bridge surgery in humans. After four and a half years at the Karolinska, he returned to his hospital in Taipei where he joined "the Swedish gang," a small group of Taiwanese doctors who had done their research studies in Sweden. He set up a lab for molecular regeneration research as well as a new clinical regeneration department and applied for permission to conduct a human trial. The problem with human surgery, however, is that in most cases the spinal cord is only crushed, rather than being neatly cut in half as in the rat experiments, and at least some fibers still cross the injury site. The bridges may do more harm than good if they destroy these valuable remaining fibers. Cheng says that when operating on people with partial, or in technical language, "incomplete" injuries, he avoids damaging the spared fibers and sometimes makes the decision to use only growth factors, not the bridges. However, the spared fibers complicate not only the surgery but also the evaluation of effectiveness: When an injury is incomplete, some degree of recovery occurs without any treatment. By mid-2001, Cheng had operated on approximately 50 people with injuries to the lower spinal cord, but the results were not yet published in scientific literature and therefore could not yet be evaluated by the scientific community.

Meanwhile, in Sweden, Professor Olson set out to explore the possibility of making the bridge surgery available to people paralyzed so long that their injury can, with certainty, be diagnosed as "complete." Such cases pose additional challenges compared with the rat study, in which the cord was repaired immediately after it was cut. In people with long-term ("chronic") injuries, a scar formed at the injury site may need to be removed, creating a larger gap that may be difficult to close by means of bridges. Moreover, immediately after the injury, nerve cells are primed for regrowth, but months or years later they may not respond as readily to the growth factor. People with chronic injuries may also develop complications, such as cysts forming at the injury site, which make the repair more challenging. Olson's laboratory, which has one of the broadest research programs in spinal cord repair, entered the twenty-first century addressing these and other aspects of chronic injury. It also sought to answer one crucial question: How did the combination bridge approach work? Did it just heal the spinal cord? Or did it restore the communication between the cord and the brain?

This question highlights the difference between rat and human regeneration experiments. Because rats are four-legged animals, they receive sensory input from their front legs while moving about the cage after a spinal cord injury. This input could be transmitted to the hind legs, causing them to move without any signals coming from the brain, thanks to so-called spinal walking, the reflexive, involuntary movements controlled by the spinal cord. In contrast, humans need their brains in order to walk, although surprisingly less so than was previously thought, as innovative rehabilitation therapies recently showed. A therapy relevant for humans would therefore have to restore a working connection between the spinal cord and the brain. For this reason, many researchers believe that before regeneration therapies are tried in humans, they must first be demonstrated to work in monkeys, whose nervous system is more similar to the human than that of rats.

One of Olson's projects at the turn of the new century is dictated by the fact that researchers cannot ask rats the same questions they would address to human subjects: What do rats feel and how do they use their brains as they recover from spinal cord injury? To answer these questions, Christian Spenger, a Swiss researcher in Olson's team, has adapted the use of functional magnetic resonance imaging (fMRI), an imaging method used to study the workings of the human brain, to the study of rats. This method, available since the early 1990s, makes it possible to see the brain in action because it detects the tiny amounts of oxygen-rich blood that flow to a brain area involved in a particular task. Applying it to rats is a huge technical challenge because of the small size of a rat's brain. Dr. Spenger's study is designed to tell researchers whether rats receiving treatment for spinal cord injury have sensation in their hind legs (if they do, their brain image will light up in bright spots), and whether a signal from their brains is sent to the hind legs when they walk (these signals would also be detected by fMRI).

Professor Olson believes that before the bridge therapy is applied to human spinal injuries it needs to be reproduced in animals by an independent team and shown to work in chronically injured rats. However, he is one of the strongest proponents of trying daring new therapies in humans, and most of the work in his laboratory is ultimately aimed at applying regeneration studies to people. The bridges, or other approaches, may come too late for the friend whose injury prompted him to go into spinal research more than 30 years ago, but since that accident Olson's outlook on achieving an effective therapy has completely changed. "For all these years I thought this was impossible, but at least I don't think so anymore," he says.

# II
# THE MANY FACES
# OF HOPE

# 10

# Hope Against Hype

Once a pursuit driven by intellectual curiosity, spinal regeneration research had turned by the first decade of the twenty-first century into an area marked by great urgency to translate scientific knowledge into a cure. After the 1996 success of the Karolinska Institute study, reports about rats and mice recovering from paralysis thanks to different approaches started coming in fast and furious. The successes of rodent studies built hopes to an all-time high, generating expectations of a treatment for spinally injured humans. "Man, to be a rat now," the spinally injured actor and director Christopher Reeve told a gathering of neuroscientists in the year 2000. "I never thought I'd be jealous of a rodent."

Hope is nothing new in spinal cord repair. Ever since people started surviving spinal cord injuries, sporadic attempts have been made to get them out of wheelchairs. But over the years, hope has gotten itself a bad name. Some people with spinal cord injury recover substantially, and charlatans touting miracle cures have occasionally taken advantage of spontaneous recoveries, attributing them to their treatments. True believers among physicians, driven by a genuine desire to help, have tried experimental procedures on people with paralysis and argued with religious fervor that their patients have been "healed." Promises of cures were invariably proven false or blown out of proportion, and *hope* became synonymous with *hype.*

What's different about hope in the early twenty-first century is that now it's official. For the first time in history, scores of prominent, respectable researchers are saying: Yes, we may be able to repair the spinal cord. The 1990s were designated by the U.S. Congress as the Decade of the Brain, but in retrospect they could also have been called the Decade of the Spinal Cord. The decade saw the advent of the first drugs for spinal cord injury. In 1990, the U.S. National Institutes of Health recommended a treatment aimed at preventing paralysis—a goal that is easier to accomplish than reversing paralysis once it sets in. The treatment, the first-ever official drug therapy for the spinal cord, consists of high doses of the steroid hormone methylprednisolone, which must be given within eight hours of the injury. The steroid is believed to reduce disability after injury by limiting trauma-related damage. Its effects are relatively small, but its impact was great: For the first time, researchers and physicians were encouraged by the medical establishment to think about spinal cord injury as treatable.

The 1990s also saw the largest-ever drug trial in spinal cord injury: a six-year, $20-million, 800-patient trial of Sygen, or GM-1, a fatty fragment of cell membranes. In the trial, sponsored by Fidia Pharmaceutical Corporation, Sygen was given within 72 hours of the injury in order to promote recovery. The drug, made from purified extracts of cow brains, had never been extensively tested in spinally injured animals, but in a rather unusual step, Fidia gambled on the substance because it was shown to have protective and growth-stimulating effects on neurons, which indicated that it could reduce nerve damage and promote regeneration. Sygen had indeed initially promoted recovery in a small trial of spinally injured people, but by mid-2001 the U.S. Food and Drug Administration (FDA) had not yet decided whether the results of the larger trial warranted approving the drug. Among Sygen's celebrity recipients have been New York Jet Dennis Byrd, who recovered the ability to walk after sustaining a partial spinal cord injury during a football game in 1992, and the gymnast Sang Lan, whose lower body has remained paralyzed since she fell headfirst during warm-ups for the 1998 Goodwill Games.

Still in clinical trials in 2001 was a third drug, called 4-aminopyridine, or 4-AP, intended to compensate for the loss of myelin sheaths by nerve fibers after injury. The drug, which can be taken in tablet form, was initially derived from coal tar but is now manufactured synthetically. It was originally applied to multiple sclerosis, but in the mid-1980s it was first used in animal studies of spinal cord injury at New York University. In the clinical trials, sponsored by Acorda Therapeutics, 4-AP is even being given to some people who have been paralyzed for years. The idea is that the drug may restore function by reviving demyelinated fibers and allowing these fibers to conduct impulses effectively. In fact, compensating for the loss of myelin nerve wrapping is probably one of the most immediately achievable goals in spinal cord repair.

As for regeneration, arguably the toughest task in spinal cord repair, during the 1990s research in this area matured, grew, and moved from the fringes to the mainstream of neuroscience. In the year 2000, the topic of the Presidential Symposium at the annual meeting of the Society for Neuroscience, a major, six-day conference that brings together some 30,000 neuroscientists from all over the world, was "Restoring Function After Spinal Cord Injury." Choices of the symposium topic reflect what's hot in neuroscience, and the rise of spinal cord injury to the top of the list was a sign of the times. No less symbolic in terms of the field's end-point goals was the inclusion among the speakers of a nonscientist who deserves much of the credit for instilling the field with hope.

Christopher Reeve has tirelessly campaigned to place spinal cord repair high on the national and international agenda since his 1995 accident. "Practically from the day he stopped moving, Reeve has not stopped moving," wrote *Time*. The image of a paralyzed

Superman touched the hearts of people around the world, helping bring in more funding and create more research projects. (In all the eight countries where I conducted research for this book, people asked if I was going to interview "the Superman.") Reeve's efforts have been all the more effective because they coincided with un-precedented advances in the science of spinal cord repair. "Had Chris's accident happened five or ten years earlier, I'm not sure his impact would have been as profound because the science wasn't ripe," says the director of research at the Christopher Reeve Paraly-sis Foundation. At the 2000 symposium, Reeve told the scientists he might be able to wait three to five years for a treatment, but not much longer than that, and called to hasten the transition from research in rats to clinical trials in humans.

The hope that Reeve promotes is not readily embraced by all people with spinal cord injury. More than 3,500 years have passed since the famous Egyptian papyrus declared spinal cord injury "a disease that cannot be treated," but in a way the statement still holds: Even though many people now survive these injuries, there is no cure for paralysis. If the Egyptian verdict is overturned within a few years, it will be a momentous accomplishment stemming from a quest that will have lasted a mere blink of an eye in the history of humanity. But for people living in expectation of a cure, every year can be an eternity.

Overblown hope, some fear, may undermine the psychological acceptance of reality required for leading a fulfilled life after paraly-sis, especially in the young and newly injured. "For 28 years I've been hearing that a cure is just a few years away," physician and political columnist Charles Krauthammer, who injured his spinal cord when he was 22 and has since been in a wheelchair, wrote in *Time* in February 2000. In his essay "Restoration, Reality and Chris-topher Reeve," Krauthammer, outraged by a prime-time commer-cial with a computer simulation showing Reeve getting out of the wheelchair, accused the actor of promoting a fantasy. "The false optimism Reeve is peddling is not just psychologically harmful, cru-elly raising hopes. The harm is practical too. The newly paralyzed young might end up emulating Reeve, spending hours on end preparing their bodies to be ready to walk the day the miracle cure comes, much like the millenarians who abandon their homes and sell their worldly goods to await the Rapture on a mountaintop." Indeed, the greater the hope, the more difficult the wait. Laboratories working on spinal cord repair are flooded with inquiries from spinally injured people eager to become research subjects or offering large sums of money in exchange for undergoing experimental therapies. In one case, a group of people with spinal cord injury, after being told that a University of California therapy that worked in paralyzed rats was not ready for human use, threatened to sue the university, a state-funded institution, for "depriving" California taxpayers of the treatment. Unmet expectations breed conspiracy theories, according to which a cure for paralysis is being withheld by manufacturers of products for the disabled or by scientists wary of losing their jobs.

But hope can also be a lifesaver. "People with spinal cord injury have one of the highest suicide rates of any population in the world," Dr. Wise Young of Rutgers University wrote in a letter to *Time*, rebutting Charles Krauthammer's views. "Most people do not commit suicide because they have had too much hope and were disappointed. Rather, they commit suicide because they have been deprived of hope." When the journalist Dennis Byrne wrote a column in the *Chicago Sun-Times* in February 2000 praising Christopher Reeve's commercial, he received so much support mail from his readers that he wrote a follow-up column, "Don't Tread on Our Hope." One of the letters he quoted there came from Penny Boyer, a spinally injured woman of Midlothian, Texas, who said she hopes for a cure even though it might be too late for her: "Without that hope, I can guarantee you I would have not survived the nine months I was in the hospital and rehab."

## From Rats to Humans

Rutgers' Dr. Young, who has written numerous essays and articles on the topic, including an editorial in *Science* entitled "Fear of Hope," believes that hope has been the major catalyst behind the current upsurge in spinal cord repair research. "Without hope, research would never get done," he says. Young's own career is one example. The Hong Kong-born neuroscientist was trained as a neurosurgeon in the 1970s at New York University, where William Windle, the one-time lonely crusader for regeneration research, had initiated a research program into spinal cord repair a decade earlier. Young was told by some of his colleagues to stay away from a hopeless field like spinal cord repair, but he ignored the advice, followed up on Dr. Windle's studies, and went on to become one of the moving spirits behind the revival of the field. He inherited a cabinet full of Piromen, Windle's failed regeneration drug, but also Windle's enthusiasm; throughout his career, he has maintained Windle's sense of urgency about making spinal cord research relevant to humans.

One of Young's major contributions to spinal cord research has been a model of animal injury, now used by numerous laboratories around the world, that closely resembles injuries in humans, in which the cord is most often bruised or crushed rather than cut across. Young also conducted some of the first animal studies on the steroid therapy for spinal injury and, together with other researchers, spent years convincing physicians to use the treatment in humans. At the W. M. Keck Center for Collaborative Neuroscience he now heads at Rutgers, his philosophy is physically tangible: In the spacious, tastefully designed laboratory, everything—cabinets, research instruments, bookshelves—is accessible from a wheelchair. While the center makes no deliberate attempt to hire spinally injured scientists, the design reflects Young's drive to make research accessible to people in wheelchairs in every meaning of the word. He travels to different countries training other scientists in the use of his animal model, which makes it possible to conduct rat studies most directly applicable to humans, and as a physician, he has been involved in several efforts to bring experimental therapies to human trials. "Access" also means talking about spinal cord research to nonscientists in lectures and at regular seminars the Rutgers center holds for spinally injured people. Moreover, in addition to supervising close to 20 research projects in his laboratory, Young spends several hours on the internet every day—or rather, every night, as he usually logs on after midnight—explaining research findings to nonscientists at *www.SCIWire.org,* a web site supported by Rutgers University. He writes articles warning people to watch out for scams, costly "cures" containing "exotic" ingredients, fanciful claims. One of his major goals, he says, is to help spinally injured people distinguish between hope and hype. Over the years, Young has become an unofficial liaison between the scientific community and the public on the topic of spinal cord injuries, and he is daily bombarded with questions about the promise of a paralysis cure: "Dr. Young, when will a cure come?" "Dr. Young, WHEN WHEN WHEN are you planning on human trials?" "Can someone please tell me WHY human trials are not being done? I'm 23 years old and supposed to be having the best years of my life. I would try anything."

Why indeed aren't all promising therapies quickly tested in humans? There is no shortage of volunteers: Many people with paralysis say they would be willing to try "anything" for a chance to be cured, regardless of the risks. Is it ethical to try risky new therapies in desperate people? Is it ethical not to try?

Dr. Young and other scientists say rushed human experiments can delay the advent of effective therapies rather than speed it up. An experiment that goes wrong can label an entire category of trials as unsafe, sending years of research down the drain and setting the field back. The tragic case of the teenager Jesse Gelsinger, who died after receiving an experimental gene therapy for a liver disorder in the fall of 1999 at the University of Pennsylvania, is one recent reminder that things can turn awry even in the most promising areas of modern medicine. And death is not the only adverse effect scientists fear. "You know what is my worst nightmare? That instead of recovery we may produce pain!" says Professor John Steeves, director of CORD, a spinal cord research center at the University of British Columbia in Vancouver. He refers to the risk that regenerating fibers in the spinal cord may create new circuits that will give rise to pain without necessarily restoring sensation or movement. Another concern is that improper wiring of regenerated circuits can derail nerve signals with disastrous results: For instance, the brain may command the body to step back from a precipice, and the legs will walk forward instead. Then there is the question of efficacy. Inadequate prior testing can cause potentially valuable therapies to be discarded as useless. Moreover, research money is limited, and it makes sense to choose the most effective therapies in animal studies before launching the more costly human trials.

Money is time. Increased funding for research can help bring treatments to humans faster, saving enormous sums of money for society. Spinal cord injury research receives relatively little funding compared with its staggering costs. In the United States, these injuries cost the nation an estimated $10 billion per year for medical and supportive care alone. (For example, in the first year after injury, health care and living expenses of a person with a high neck injury are close to $550,000.) For pharmaceutical companies, spinal cord injury is not a major draw. It costs on average $500 million to take one medicine "from bench to bedside," and Dr. Young has calculated that in the case of several drugs for neurological conditions, the process took on average more than 11 years. The number of spinal cord injuries is relatively small compared with this investment, and unless the treatment can be applied to other neurological disorders, most companies balk at the costs. In 2000, according to a typical estimate, a total of $100 million was spent on spinal cord injury research in the United States, including federal funding, privately raised funds, and industry investment. This means, Young points out, that for every dollar of care costs, only one cent was being spent on research. Other scientists voice similar views. "If you get cancer, you are either cured or you die, and both are not too expensive," says Lars Olson of the Karolinska Institute. "But if you have a spinal cord injury or Parkinson's disease, you live a long life, with many problems and many costs to society. I'm not saying that cancer research gets too much, only that neuroscientists have not been good at explaining the societal benefits of their work." The greater the investment in research, the more likely it is to pay off. "It's a risk venture. You're investing money in a high-risk field. But if I invest money in 20 therapies and one works, it was worth the

other 19," says Murray Goldstein, former director of the National
Institute of Neurological Disorders and Stroke.

## Electrifying Hope

One could argue that testing regeneration therapies in humans is
nothing new. Pages can be filled with accounts of therapies already
tried in spinally injured humans, particularly in countries where
human experiments are not strictly regulated by authorities. The
problem is that when the therapies are tried in a haphazard, uncon-
trolled manner, researchers can easily be fooled by a complex and
unpredictable organ like the spinal cord. Unlike the uniform injuries created in laboratory animals, each human injury is unique,
which makes comparing treatment results in different people extremely problematic. Moreover, people who receive no treatment
often spontaneously recover some sensation and movement after
spinal cord injury, and this recovery muddles the evaluation of experimental therapies. Whenever a person's condition improves, the
question arises: What would have happened if he or she had not
received the treatment? Would they have recovered anyway?

The history of spinal cord regeneration is rife with human experiments in which these questions were never properly answered.
One famous—or, rather, infamous—example is omentum surgery,
an operation in which a portion of the omentum, blood-vessel-rich
tissue from the abdominal cavity, is placed over the injury site in the
spinal cord. The idea, never proven definitively, is that this tissue
releases substances that stimulate nerve fibers to regrow. Hundreds
of such operations are said to have been performed on spinally injured people in China, Japan, and Mexico, and there are unconfirmed reports of individuals recovering some function after the
surgery. But a clinical trial of the procedure launched at a Boston
hospital in the 1990s ended in a mess: It was stopped by Massachusetts authorities because of concerns over the operation's safety and
efficacy; several participants claimed the trial was mismanaged and
filed legal suits against the hospital. There is no firm evidence that
omentum surgery doesn't work; but neither is there firm evidence

that it does. To many neuroscientists, its use in humans is a prime example of how a procedure taken to clinical trials before its scientific basis is established ultimately harms  the entire field of spinal cord repair.

The new wave of human regeneration trials of spinal cord injury—the ones already started, like the Israeli trial in which the American Melissa Holley became the first participant, and those expected to begin in the first decade of the twenty-first century—are different from past attempts. Most will be approved by regulating bodies such as the U.S. FDA, which makes sure they are anchored in meaningful scientific evidence. But no less important, an approval also means the trials are designed to provide crucial answers about what works in spinal cord repair and what doesn't. One way to ensure such answers is to include only people with long-standing, "chronic" spinal cord injuries who are no longer expected to improve. Another possible category is that of newly spinally injured people with "complete" injuries, who have no movement or sensation below the injury site. Major recovery is not expected to take place after such injuries, and if it occurs in a significant number of people, it can be attributed to the treatment.

One latest example of a controversial spinal cord therapy that is being put to a test in a rigorous, FDA-approved trial is electrical stimulation of nerve regeneration. The idea behind it is appealing: Back in the 1920s, it was suggested that an electrical field can enhance or direct nerve growth, and subsequent studies showed that in a laboratory dish, nerve fibers tend to grow toward the negative pole. Work in this area continued in several countries, and in the former Soviet Union, where the tendency was to try new therapies in humans sooner than in the West, human experiments were launched as early as the 1970s. Moscow neurosurgeon Arkady Livshits, after experimenting for five years on dogs, implanted his first electrical device for stimulating the injured spinal cord into a human being in 1973. Livshits eventually placed such devices, custom-made for him in a space-rocket factory, in 10 spinally injured people and in 1977 reported in a Russian-language journal that his patients had improved.

Meanwhile, enthusiasts elsewhere pursued this line of research in animal studies. In the United States, one of the leaders in this field, Richard Borgens, used electrical stimulation to produce nerve regeneration in frogs and lampreys, and in the early 1980s, he moved on to guinea pigs and dogs. But attractive as the idea was, electrical stimulation never became a major direction in spinal cord regeneration; studies in this area have not been widely cited in scientific reviews of the regeneration field. Dr. Borgens, however, found a unique and creative way to continue his research; working at Purdue University's school of veterinary medicine, he started treating dogs naturally paralyzed in accidents. Together with his Purdue colleagues, he developed an implantable electrical device for stimulating the spinal cord after injury, and during the 1990s implanted such devices in more than 100 paralyzed dogs. In the fall of 2000, accumulated data allowed Borgens, together with physicians from Indiana University School of Medicine, to obtain permission from the FDA to test the implantable device in 10 newly injured humans with spinal cord injuries. If the approach is shown to be safe, the researchers have the option to request permission from the FDA for another 10 cases. The design of the trial promises that the uncertainty over electrical stimulation may now be resolved: The trial includes only people with "complete" injuries of the spinal cord.

## One Hell of a Job

Many leaders of the neuroscience community today are willing to go on record saying they are hopeful about the prospects for successful regeneration and repair of the spinal cord in humans, even though they are usually quick to qualify their statements with caveats. In an October 2000 review article on central nervous regeneration in *Nature*, prominent neuroscientist Fred Gage, of the Salk Institute for Biological Studies in La Jolla, California, wrote that "the first rational, functional therapy for regeneration, probably for spinal cord injury, may be in the clinic just 10 years from now." But the scientists immediately point out the difficulties.

In most animal studies regeneration in the spinal cord is still rather modest, about 10 percent of the damaged fibers, and scientists are striving for more. Another challenge is to have the fibers regrow over long distances. In some cases, even short-distance growth can bring tremendous relief to a person with paralysis: An inch of nerve growth in the neck area, for instance, can make a difference between breathing independently or being hooked to a ventilator. But to regain full control of the leg muscles it is possible that some of the human nerve fibers may have to grow a foot or longer, and no one knows if they can. Finally, a major question hangs over the field: Will the regenerating fibers establish effective, working connections and circuits? This question cannot be definitively answered in rats: Even when rats recover from paralysis, their movement may originate in neurons inside the spinal cord. Researchers cannot be sure that regeneration in rats restores the kind of a complex nervous architecture that ensures communication with the brain, which is necessary for restoring function in humans. Neither can the answer to the question be based on regeneration of peripheral nerves: Nerve circuits in a finger or even a hand do not have to be restored as precisely as the ones in the spinal cord. When nerve cells "grow up" together in the fetal cord, they form circuits in which they learn to work together. Will the same happen during regeneration in humans? Nobody knows if this can be achieved, or how to make sure the fibers are reattached in a meaningful manner.

There are many other unknowns and one certainty: The "enigma of paralysis" is vastly more complex than regeneration pioneers ever thought. A simple, one-bang solution for paralysis is unlikely to come about; the injury is too catastrophic to be repaired by one therapy. Rather, spinal cord repair hinges on a convergence of different approaches. The complexity, however, has its bright side; it has brought with it new avenues of hope from different areas of neuroscience and from a wealth of unexpected directions. Dozens of therapies now hold potential for promoting regeneration in the central nervous system, and the best ones will have to be combined to promote recovery.

The picture of hope for spinal cord injury at the turn of the new millenium is a gigantic mosaic that is being rapidly filled in. Now we can watch major pieces of this mosaic move into place, including the therapies already being tested in humans or expected to reach humans soon: enrolling natural mechanisms in repair, using growth-stimulating molecules originally discovered in a World War II drama, "educating" the spinal cord after injury. Regeneration research is being extended even to long-standing spinal cord injuries, and prosthetic devices may offer interim solutions or help people with the most severe disabilities. In the early twenty-first century, spinal cord repair is a busy, competitive field with dozens of research teams around the world racing to produce an effective therapy for spinal cord injury. The odds that their efforts will one day result in a victory over paralysis may be summed up by a statement on nerve regeneration in a recent editorial in *Nature Medicine*: "It looks like one hell of a job—but one that might well be possible."

# 11

# Healing from Within

One new way of tackling spinal cord repair is to declare a cease-fire with Mother Nature: Rather than eliminating one hurdle to regeneration or another, scientists attempt to enlist the help of different natural mechanisms to coax the spinal cord into repairing itself. But perhaps the most powerful natural repair kit, the immune system, is also the one that presents a quandary. The immune system is the body's resident healer; it mends and repairs most tissues, yet for the brain and spinal cord its action generally spells nothing but trouble. Why is this so? The Israeli scientist Michal Schwartz believes she has the answer; and in a creative new approach, she claims to have tamed the immune response. It was Schwartz's immune therapy for the spinal cord that the American teenager Melissa Holley received.

Schwartz, a slender mother of four with a halo of light-chestnut curls and a seemingly limitless supply of fiery energy, likens her immune therapies, the result of 20 years of research, to "a really long pregnancy that has finally led to a baby." She had decided to become a scientist at age 11, while still in elementary school, and vividly remembers visiting a children's workshop on the lush, green island of a campus of the Weizmann Institute of Science, where she now works. The Institute seemed like a magical, impenetrable world where the mysteries of life were being probed. Schwartz studied chemistry and physics in Jerusalem, then came to follow a Ph.D. program in immunology at the Weizmann Institute in Rehovot, a small town a half hour's drive from the Mediterranean coast. Combining long hour's in the lab with raising a family was made easier by the active interest that her children took in their mother's career. Propped up on the window of Schwartz's office is a rag doll imprinted with the words "To the nicest professor," a present she got from her children, then still small, when she received tenure at the Institute. During a postdoctoral stint in the United States, Schwartz had come across a topic that for the first time truly fascinated her: nerve regeneration. She switched to neuroscience and, upon returning to Weizmann, joined the department of neurobiology but retained an interest in the immune system.

Back in the 1950s, William Windle's failed wonder drug Piromen had pointed toward a possible link between immune responses and regeneration. Piromen's mechanism of action was never revealed, but according to one theory, it somehow provoked nerve growth by affecting the immune system (at least the fever it produced hinted at this). In subsequent decades, the notion of the immune system's role hovered over regeneration studies, but the prevalent view was that the immune response—and, particularly, inflammation—was bad for the nerves. When Schwartz embarked on nerve regeneration studies in the early 1980s, she was inspired by Dr. Windle's writings about the immune system. Windle, with his antidogmatic stance, never claimed the immune response was good for regeneration, but he argued it was an area worth studying. Schwartz, who describes herself as a neuroimmunologist, an expert on the cross talk between the immune and nervous systems, wanted to shed light on the tantalizing hints about the immune involvement in nerve growth.

It is hardly surprising that the immune system was long considered bad news for the brain and spinal cord. Harmful inflammation caused by the immune response is one of the characteristic features of numerous diseases of the nervous system, such as meningitis and multiple sclerosis. According to the latest findings, inflammation also plays a role in degenerative brain disorders such as Alzheimer's disease. In trauma to the brain and spinal cord, an exaggerated immune reaction creates an additional, mechanical problem. Injury breaks the blood–brain barrier, the protective shield of the central nervous system, and some inflammatory cells rush to the injury site, aggravating swelling. "A bit of swelling in your arm is not a big deal. There the influx of inflammatory cells is good for removing debris, but in the brain the swelling has nowhere to go, it pushes on tissues and then you die. That is why one of the first things in brain trauma is to monitor intracranial pressure," says Professor Hugh Perry, of the University of Southampton in the United Kingdom, who has conducted extensive research on inflammation in the central nervous system.

Our everyday experience also tells us that inflammation is a bad thing; inflamed skin or an inflammatory reaction in internal organs signals discomfort and disease. But our everyday experience is misleading. Inflammation arises in body tissues for a good reason; it is an immune system's attempt to overcome infection or damage, and it becomes a problem only when the immune system overreacts or when its reaction persists too long. Therefore, even if inflammation is the hallmark of many brain and spinal cord disorders, pinning it down as the original troublemaker may be a mistake. Just as fever is not the disease itself but part of the body's attempt to fight disease, so inflammation is a normal attempt to restore balance. For example, such inflammatory reactions as redness and swelling caused by a cut in the skin are an essential part of wound healing. Professor Schwartz regards nerve regeneration as a special case of wound healing. A cut or injury in a nerve, in her mind, is just another kind of wound that could be repaired by a controlled dose of the immune system's inflammatory response.

The inflammation idea came from research on fish. Schwartz and her team studied the optic nerve, part of the central nervous system, which regenerates in fish but not in mammals. She compared fish and mammals in their response to optic nerve injury in the hope of developing a strategy for effectively regenerating mammalian nerves. The studies focused on two unlikely relatives, the carp and its pretty cousin, the goldfish. (Despite the differences in their appearance, both freshwater species belong to the Cyprinidae family of fish.) The carp were particularly useful: Because of their large size, they yielded abundant material for research. Before performing the experiments, scientists caught them by the head and tail and dipped them in a bucket filled with diluted sleeping pills. (Because the plain-tasting carp are used to prepare a staple of Jewish cuisine, gefilte fish, the humor of having a national delicacy swimming in the basement was not wasted on people working in the building. They jokingly regretted that the carp were too filled with sleeping pills to be grilled for a meal.) The scientists crushed the carp optic nerves, waited for regeneration to set in, then squeezed the regenerating nerves and applied the resulting soup of chemicals to the optic nerve of rabbits.

The fish nerve extract produced regrowth in the rabbit optic nerves, but the study, reported in 1985 in *Science*, stumbled into the same roadblock that had impeded many other attempts to transfer fish regeneration findings to mammals: What exactly was stimulating the regrowth? A squeezed fish nerve was a legitimate research tool, but without understanding what exactly was in the "fish juice," as the scientists called it, the extract could never be applied to humans.

The search for growth-stimulating substances in regenerating fish nerves pointed toward inflammation. Schwartz and her colleagues began to compare the inflammatory response in situations where regeneration does occur with those where it does not. In particular, the scientists focused on the differences in the immune response to injury between peripheral and central nerves. By the mid-1990s, Schwartz had developed the hypothesis that one of the main differences between the inflammatory response in injured central and peripheral nerves lies in the nature and number of immune cells that arrive on the scene.

Until recently it was thought that immune mechanisms had no place at all in the healthy brain and spinal cord. The central nervous system is guarded by the so-called blood–brain barrier, which creates a chemical and physical moat around the brain and spinal cord. The barrier shields the central nervous tissue from infectious organisms and from unwanted substances; most large molecules cannot cross from the blood to the brain, and neither can immune cells. The new field of neuroimmunology, however, has revealed that the blood–brain barrier is more leaky than previously thought and that there is much more interaction between the immune and nervous systems than was believed in the past. One fascinating aspect of this cross talk is the link between emotions and disease: The state of mind apparently influences health via the signals exchanged by the brain and the immune system. Still, the brain and spinal cord allow immune cells more limited access than do other body tissues; its restricted accessibility gives the central nervous system a so-called privileged status.

Schwartz believes this state of affairs, known as "the immune privilege," is a by-product of evolution. When most tissues are damaged, the immune system bursts on the scene with all the subtlety of a vacuum cleaner, removing debris and killing damaged or infected cells before initiating the healing process. Fish and amphibians, with their relatively simple brains, can afford the damage, but not mammals: If such a brute-force helper were given free access to the complex mammalian brain, it would wipe out the intricate circuitry built up throughout an individual's life. In severe injury, Professor Schwartz believes, this privileged status of the mammalian central nervous system backfires: While most other tissues are successfully healed by the immune system, the brain and spinal cord are not.

In other words, lack of regeneration is the price that mammals paid for having complex and dynamic brains. "There seems to have been an evolutionary trade-off," Schwartz says. "Higher animals protected their central nervous system from invasion by the immune system but paid the price of losing their ability to regenerate injured nerves." Schwartz's approach is to temporarily strip the spinal cord of the "privilege" for its own good. She wants to allow the body's own healing mechanisms to perform two related tasks: One of her experimental therapies is aimed at regenerating fibers in the injured spinal cord, while the purpose of the other is to limit the cascade of damage that follows nerve injury.

## The Big Eaters

In most body tissues, the injury site fills up within the first 24 hours with large immune cells called macrophages. True to their name, which literally means "big eaters," they perform the role of all-purpose cleaners, swallowing anything that should not be there and preparing the affected area for healing. Their taste is eclectic: In infection, they gobble up bacteria or other foreign organisms; in the case of mechanical trauma, their meal consists of dead cells and debris. Once they have consumed something unsavory, the macrophages become "activated." They release growth factors and other substances needed for tissue repair. In regeneration of peripheral nerves, macrophages invade the injury site and help clean up myelin fragments and other debris that may interfere with regrowth of nerves. They then unleash chemicals that directly or indirectly stimulate the growth of nerves.

Schwartz extended her studies to parts of the central nervous system other than the optic nerve and found that in the spinal cord, macrophages arrive to the injury site in small numbers and fail to become adequately activated in a way needed for stimulating regrowth. Together with graduate student Orly Lazarov-Spiegler and other team members, she used the finding to develop a strategy for spinal cord repair: Take a few macrophages, incubate them in the presence of damaged peripheral nerves to "activate" their healing touch, then inject them into the damaged spinal cord. Now working with rats, the scientists performed a complete cut of the spinal cord, including all its surrounding membranes, and bridged the gap with organic glue to provide the nerves with a surface for growth.

After giving the macrophage treatment, Schwartz and her team watched their rats drag their hind legs around the cage for six long weeks. Then, in the seventh week, some of the rats started using their legs and attempted to stand upon them; by the twelfth week, 75 percent of the treated rats were using their legs to some extent. The excitement spilled beyond the laboratory and engulfed Schwartz's entire family. Her oldest daughter, Orit, was particularly impressed. She was in medical school and all her textbooks still said that effective regeneration never takes place in the spinal cord. But then doubt set in: Perhaps the spinal cords of the rats had simply not been properly cut? To dispel all doubt, Schwartz asked a visiting scientist from Sweden's Karolinska Institute to help her repeat the experiment. The Swedish scientist performed the cutting of the cord, and most treated rats again showed signs of recovery. They were far from walking normally, but some of them could move their legs and place their weight upon them—on a scale of 0 to 21, where 21 is normal locomotion, they scored on average 7; in contrast, the untreated rats remained paralyzed and scored on average 1.

Schwartz's approach is a rather drastic departure from conventional views on inflammation in the central nervous system. Some scientists believe inflammation is always bad for the nerves; others are wondering whether some aspects of inflammation are beneficial but adopt a cautious "wait and see" stand on the macrophage therapy. Professor Hugh Perry, whose work on macrophages had provided Schwartz with some initial elements for her theory, says the macrophage story is complicated. "I think it is not clear whether macrophages are good or bad for injured central nerves. In many other circumstances, they clearly don't do neurons any good. There is no doubt macrophages clean up debris in peripheral nerves, but the question is: Is it essential?"

Schwartz's macrophage therapy, however, has one major advantage over many other regeneration approaches: It makes use of the person's own cells. Transplanting the patient's own tissues is considered safer than introducing a foreign substance into the body. That was what helped the macrophage therapy make the crucial transition from lab to life at record speed. Less than two years after the rat recovery study was published in 1998, Proneuron Biotechnologies, an American–Israeli company created to develop clinical applications of Schwartz's research, obtained permission from the Food and Drug Administration (FDA) in the United States to conduct a human trial.

The macrophage trial is a harbinger of others to come. When the FDA approved the testing of spinal cord therapies in the past, in most cases regeneration was a possibility at best. But the macrophage treatment, as well as several other new therapies to be tested in humans in the first decade of the twenty-first century, are primarily and explicitly aimed at regenerating the spinal cord.

Ever since reports on the trial reached the media, Proneuron has been deluged with letters, calls, and e-mails from paralyzed people all over the world, some of them offering to pour all their savings into the company in exchange for receiving the experimental therapy. However, only people with new injuries are admitted because in the rat study, the macrophages were shown to be effective if given within two weeks of injury. Another criterion for taking part in the study concerns safety: The spinal cord must be damaged lower than the C5 vertebra, around shoulder level, so that any potential complications would not affect the higher spinal nerves responsible for breathing and other vital functions. No less important, only people with complete injuries, the ones whose spinal cord is so badly damaged that they have no movement or feeling below injury level, are admitted. The completeness of the injury is precisely what makes it possible to evaluate the results of the therapy: People with complete injuries typically regain no function in their lower bodies. Therefore, should a substantial recovery occur in a significant number of cases, it can be attributed to the treatment.

The first few locally injured individuals who were willing to receive the Weizmann Institute therapy had to be turned away by the Israeli doctors running the trial because they did not meet the strict inclusion criteria set by the FDA. Melissa Holley, whose accident occurred in Colorado on June 25, 2000, was the first spinally injured person to meet all the requirements. Her flight to Israel in a private ambulance jet was dictated by the "window of opportunity" suggested by the rat studies. Had she waited until she could take a regular flight, more than two weeks would have passed since her accident and she would no longer qualify for the trial. The treatment she received consisted of an operation in which a neurosurgeon opened the membranes of Melissa's spinal cord and injected several million macrophages—removed from her blood days earlier and "activated" for repair—into the injury site.

Doctors told Melissa she might have to wait up to a year to know if the treatment worked. Regeneration therapies are slow to produce results because at least some of the cut or damaged nerve fibers must regrow to their previous length. In rats, nerve fibers grow at about the same pace as hair, approximately 0.5 to 1 millimeter per day, and they do not necessarily grow in a straight line; some wander through tissue and may get lost. That is why even in these relatively small animals, the process takes weeks. Human beings are larger, which means that some of the regenerating fibers must cover long distances—several inches, depending on the level of the injury and the person's height. Nobody knows how fast human neurons regenerate, but even if the speed is the same as in rats, the process can easily take many months. Moreover, establishment of new nerve connections, should it take place, can probably also be a time-consuming process.

As in any clinical trial, the therapy is tested in several phases. In the first several subjects, of which Melissa was the very first, the focus is on evaluating the safety of the treatment. If there are no negative side effects, the trial proceeds to its next phase, the evaluation of efficacy, in which physicians seek to optimize the way the treatment is given. Results on the efficacy of the macrophage therapy were not expected before the year 2002.

While Melissa was recovering back in the United States, several other people, all of them with a completely cut or crushed spinal cord, underwent the immune-based regeneration procedure in Israel. Meanwhile, in their laboratory at the Weizmann Institute, Professor Schwartz and her team were enlisting the healing powers of the immune system in performing an additional task: reducing the secondary damage that accompanies injuries to the spinal cord.

## An Ounce of Protection

In most injuries, the spinal cord is not completely cut; it is only bruised or compressed. The damage, however, continues to spread from the injury site for hours, days, or even weeks, with chemicals spilling from damaged cells and killing tissues that were spared by the original trauma. Some people can move their body immediately after the injury but are soon left paralyzed by the secondary damage. If the posttraumatic demolition process can be effectively contained, injuries that now completely destroy the functioning of the spinal cord may result in only partial paralysis. Then, many spinally injured people may not need a regeneration treatment, and even if they do, regeneration researchers will have a much easier time repairing their injuries.

Barbara Turnbull, the Canadian journalist who was spinally injured in a late-night armed robbery of a store in 1983, lay stranded in a pool of blood for 15 minutes before a passerby called for medical help. When ambulance attendants gave Turnbull first aid, she was still able to move her legs and squeeze her hands on command. But by the time she reached the hospital, she could no longer move anything and has remained fully paralyzed from the neck down. Turnbull's injury occurred seven years before the first (and so far, the only) therapy suggested to reduce secondary damage— megadoses of the steroid methylprednisolone—was recommended for clinical use. Now many other neuroprotective therapies, including that of Schwartz, are under study.

Secondary damage occurs because the mechanical injury has the effect of a bulldozer crushing a chemical factory; chemicals that were safely stored in cells spill out and damage adjacent tissues. Swelling compresses the cord and kills more neurons. Blood supply is interrupted, causing tissues to die of oxygen starvation. This ripple effect can sometimes be more harmful than the trauma itself, destroying fibers that could have allowed the person to retain or regain substantial function.

The neuroprotective approach proposed by Professor Schwartz, just like her macrophage therapy, sounds paradoxical: She seeks to contain secondary posttraumatic damage by employing the immune system's most vicious weapon, autoimmunity. Traditionally, autoimmunity, the appearance of proteins and cells that can interact with the body's own tissues, has been viewed as the immune system's tragic mistake. When these proteins or cells attack the body, they cause autoimmune diseases such as multiple sclerosis, juvenile diabetes, or rheumatoid arthritis. In multiple sclerosis, for example, autoimmune T cells destroy the protective myelin sheath of nerves.

Yet Schwartz believes autoimmune T cells are not born villains; in fact they appear to possess healing powers. This conclusion emerged from a serendipitous discovery. In a study conducted with Weizmann Institute's Professor Irun Cohen, a world expert on autoimmunity, Schwartz set out to use T cells as vehicles to deliver therapeutic genes to injured nerves. But then, based on initial observations, Schwartz began to suspect that T cells *themselves* had a protective effect on the nerves. Scientists in Schwartz's laboratory found this uncanny because it is well known that in different circumstances, T cells are known to cause disease, but when team member Dr. Eti Yoles pursued the experiments, she found that T cells indeed protected bruised fibers in the optic nerve and spinal cord: A single injection of T cells within a week after the injury saved a significant number of fibers (treated animals had three to five times more spared fibers than the ones who received no treatment) and produced recovery of function in spinally injured rats. The T cells were specially pretreated to make sure they caused no multiple sclerosis. Still, how could the same cell heal tissues in some circumstances and trigger disease in others?

The proposed answer calls into question the conventional wisdom on autoimmunity. In fact, Professor Cohen had been saying for years that autoimmunity, despite all the suffering it inflicts on millions of people, isn't intrinsically bad. Contrary to accepted views, he argued, autoimmunity is present in healthy organisms and causes disease only in exceptional cases. Now the findings in Schwartz's laboratory have led to an even more radical theory: Autoimmunity may actually have come into being for a reason: to protect and repair tissues after trauma.

According to this theory, the road to autoimmune diseases is paved with the immune system's good intentions. Autoimmune T cells are not produced by mistake, their job is to maintain body tissues and repair minor damage on an ongoing basis. This maintenance mechanism apparently operates quietly throughout a person's life, offsetting the wear and tear of day-to-day living, and makes itself painfully known only when it gets out of control, leading to autoimmune disease. The theory sheds an intriguing light on the link between autoimmune disease and trauma. Ailments like multiple sclerosis or arthritis are sometimes triggered by mechanical injury; a person bruises a shoulder or knee, and later develops arthritis in those joints. The disease may be the doing of an unruly autoimmne response that was initially called upon to contain the damage.

Schwartz believes that the access of T cells to the spinal cord may be restricted, as in the case of macrophages, because of the central nervous system's "privileged" status. And as in the case of macrophage therapy, she says her T-cell method is none other than an attempt at boosting an underused natural healing mechanism— obviously, while taking great care to avoid causing an autoimmune disease. The T-cell approach led to the development of a therapeutic "vaccination." A vaccine's job is usually to prime the immune system for protecting the body against outside invaders, such as infectious organisms; in Schwartz's method, an injection of autoimmune T cells after injury mobilizes the immune system to protect nerve tissue against self-destruction, the damaging cascade unleashed by the body itself in the aftermath of mechanical trauma. Following the success of this research, Schwartz's team has synthesized protein fragments, or peptides, that eliminate the need for the more complicated T cell injections; in rat studies, when the peptides are introduced into the bloodstream, they prompt the rodents' own resident T cells to exert a protective effect on injured nerves, without the risk of causing autoimmune disease. The peptides, currently being developed for application in humans, may make it possible to "vaccinate" a spinally injured person against secondary damage to the cord's nerve fibers during the week following the injury.

Schwartz's neuroprotective "vaccine" is but one attempt to minimize postinjury damage; dozens of other approaches are being developed in laboratories around the world. Not only is this research a vital part of spinal cord repair, it has broad implications for a host of neurological disorders, such as stroke and glaucoma. Findings are traded back and forth among researchers working on spinal cord injuries and these diseases.

As for regeneration and repair of the spinal cord, scientists all over the world are tapping the potential of a variety of cells, not necessarily those belonging to the immune system. Some of these approaches are producing promising results in experimental animals and may soon be tested in humans.

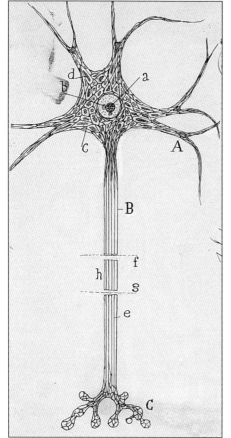

Ramón y Cajal (*above left*), one of the founders of modern neuroscience. The Spanish researcher was co-recipient of the 1906 Nobel Prize in physiology or medicine for studies on the structure of the nervous system.

Francisco Tello (*above right*), Cajal's favorite disciple and later head of the Cajal Institute in Madrid. Tello's 1911 experiments planted the seeds for the current revival of spinal cord regeneration research.

Drawing of a nerve cell by Ramón y Cajal. The tree-like branches (A), extending from the cell body, are dendrites; the long nerve fiber (B) is the axon. Cajal, who as a child dreamed of becoming a painter, made thousands of drawings of the nervous system, some of which are still used as scientific illustrations.

*Plate 1*

William Windle, who began working on spinal cord regeneration at the University of Pennsylvania in 1949, was a lonely crusader in the field for more than 20 years.

Melissa Holley. Spinally injured in a car crash in Colorado, she was flown halfway around the globe to receive an experimental regeneration therapy in July 2000. The research that led to this therapy was inspired by Windle's writings.

Participants of the first conference in history on regeneration in the central nervous system, organized by Windle at the NIH in May 1954. From left to right: Leslie Freeman, William Windle, Roger Sperry, William Chambers, Ralph Gerard, unknown, Jean Piatt, Rita Levy-Montalcini, P. Glees, H. Hoffman, unknown, Jae Littrell, unknown, Edward Stuart, unknown, Theodore Koppanyi, Chan-nao Liu, Paul Weiss, unknown, Carmine Clemente, R. Lorente de Nó, Donald Duncan, Donald Scott, Louis Flexner, Fred Stone (NIH), Grayson McCouch.

*Plate 2*

William Chambers (*left*) and Chan-nao Liu (*far right*), pioneers of spinal cord plasticity research. Their landmark 1958 study at the University of Pennsylvania remained controversial for nearly 30 years. Shown here with their wives, 1980.

Albert Aguayo, professor of neurology and later head of neuroscience research at McGill University's Montreal General Hospital. In the early 1980s, his team proved that spinal cord neurons could be regrown after injury.

Susan Keirstead, Yutaka Fukuda (*center*), and Michael Rasminsky (*right*), 1989. On the monitor, the good luck Japanese doll that "oversaw" their McGill University experiment, in which a regenerated optic nerve was shown to reestablish working connections.

*Plate 3*

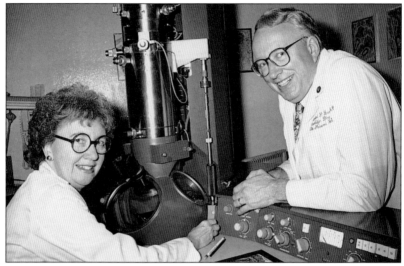

Husband-and-wife team, Mary and Richard Bunge, at an electron microscope shortly after joining the Miami Project to Cure Paralysis in 1989.

Martin Schwab of the University of Zurich (*left*) receiving the first Christopher Reeve Research Medal in 1996 at the University of California, Irvine, with Reeve and his wife Dana.

*Plate 4*

Rita Levi-Montalcini, the Italian scientist awarded the 1986 Nobel Prize in physiology or medicine for discovering the first molecule known to stimulate the growth of nerves. This picture was taken around the time of the Prize.

Sir Ludwig Guttmann, founder of the noted spinal cord injury unit at Stoke Mandeville Hospital near London. (From the 1964 portrait by Sir James Gunn.)

Roger Sperry, awarded the 1981 Nobel Prize for studies on the split brain conducted at the California Institute of Technology. In earlier research, Sperry made monumental contributions to clarifying how neuronal connections are established in development and regeneration. This picture was taken in the late 1960s.

*Plate 5*

Henrich Cheng of the Veterans General Hospital in Taipei, Taiwan (*left*), and Lars Olson of Sweden's Karolinska Institute (*right*); their 1996 study produced the first convincing evidence of recovery in rats with completely cut spinal cords.

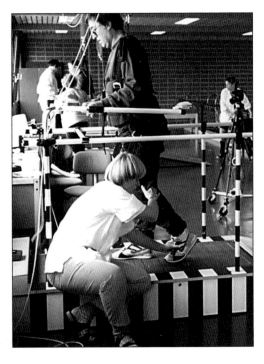

Undergoing treadmill (Laufband) training at a rehabilitation clinic in Karlsbad-Langensteinbach, near Karlsruhe, Germany. This innovative training allows some people with partial paralysis to walk again.

*Plate 6*

Wise Young, head of the W. M. Keck Center for Collaborative Neuroscience at Rutgers University, a veteran spinal cord researcher and tireless popularizer of science.

New York Jets player Dennis Byrd collided with a teammate during a 1992 football game (*below*), crushing his spinal cord at neck level. Despite initial paralysis, Byrd eventually recovered (*left*). His recovery may have been helped by methylprednisolone, still the only drug officially recommended for limiting the consequences of spinal cord injury. He also received the experimental drug Sygen.

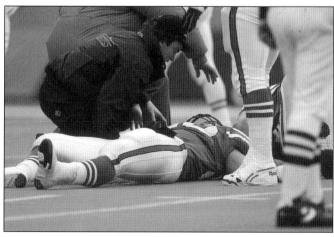

*Plate 7*

Michal Schwartz of Israel's Weizmann Institute of Science developed the immune therapy for spinal cord injury that was administered to Melissa Holley in 2000.

Results of post-injury "vaccination" developed by Schwartz and colleagues to limit the spread of damage after spinal cord trauma. Shown here, rat spinal cords five months after injury. The cords of untreated rats (*top*) have significantly fewer spared nerve fibers compared with those of treated rats (*middle*). At the bottom, an uninjured rat spinal cord.

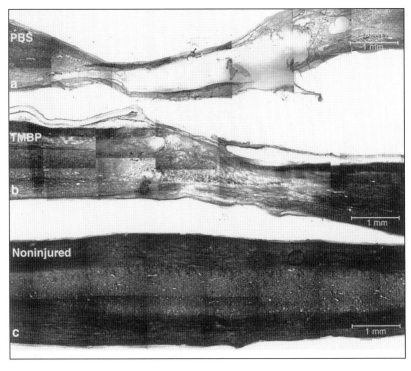

*Plate 8*

# 12

# Mighty Cells

Body cells know more about tissue repair than any scientist in the world. It is therefore little wonder that cell therapies are increasingly used in many areas of medicine. Having been around for millions of years, cells are capable of turning their genes on and off, moving to desired locations, and releasing a host of chemicals that stimulate repair and growth. Why not inject appropriate cells into the injured spinal cord and let them work their magic?

In theory, the cells may be able to furnish a combination therapy for the spinal cord even before scientists understand the repair process in all its details. They can bridge the gap created by the injury and release the right substances needed for regeneration, at the right times. And, no less important, certain cell types may be able to perform an additional crucial task: replace cells killed by the trauma. In some injuries, body function cannot be restored without cell replacement. For example, if all spinal cord motor neurons that send signals to the leg muscles are killed, no amount of regeneration of cord fibers will restore leg movement unless these neurons are somehow compensated for or replaced.

Different cells are being considered for transplantation into the spinal cord, but none get as much attention as the new superstars of biomedical research, the stem cells. These are the uncommitted cells that make up the embryo in its first week of life, before they begin to mature into different types of tissue. Because of their versatility, they promise to give rise to the ultimate cell therapy: replacing damaged areas of the body with custom-grown tissue. "Stem cells may one day be used to treat human diseases in all sorts of ways, from repairing damaged nerves to growing new hearts and livers in the laboratory; enthusiasts envision a whole catalog of replacement parts," wrote *Science,* having named stem cell research Breakthrough of the Year for 1999.

Stem cell research took off when, in 1998, two company-funded scientific teams accomplished the long-sought feat of "prolonging the moment of cellular youth," as *Science* put it: Embryonic stem cells were kept dividing in a laboratory dish while staying forever immature (without special measures, stem cells quickly disappear because they mature into different tissue types). Since then, scientists all over the world have rushed to make the cells deliver on their promise. Almost every month brings a new breakthrough in stem cell research, and scientists have already used stem cells to create more than a dozen different types of tissue in the test tube. That is why embryonic stem cells are called *pluripotent,* from the Latin for "many capabilities."

The use of human embryonic stem cells, like the use of cells from a fetus, is problematic because they are derived from human embryos, which entangles them in  the controversy over abortion. The use of "immortal" stem cells, which can be grown in the laboratory forever, may resolve the problem by eliminating the need for embryos. Another hope to avoid controversy arises from the presence of an "inner child" in every adult, small reservoirs of stem cells lurking in different parts of the body. These cells are more "committed" than the ones in the embryo; they are on their way to maturing into specific tissue: Intestinal stem cells regenerate the lining of the gut, skin stem cells make new skin, bone marrow cells produce new blood. However, researchers recently found that adult stem cells can be persuaded to alter the course of their careers. "In defiance of decades of accepted wisdom," wrote *Science*, "researchers in 1999 found that stem cells from adults retain the youthful ability to become several different kinds of tissues: Brain cells can become blood cells, and cells from bone marrow can become liver." Thus, stem cells for spinal cord repair may come either from a bank of perpetuated embryonic cells or from the person's own stem cell reservoir.

A third possibility would be to stimulate the growth of stem cells already present in the central nervous system. In fact, recent stem cell research has toppled the old dogma that the human brain never generates new cells in adulthood. Researchers had already reported in the 1980s that new nerve cells were born every year during the mating season in the brains of adult canaries, when the birds learned new songs, but scientists had been saying, oh, well, these are only birds. Then, in the 1990s, several research teams demonstrated that new neurons can be born in the brains of rodents, tree shrews, and marmosets. And 1998 brought an astounding discovery: Researchers revealed that the mature human brain routinely spawns new neurons in at least one site, the hippocampus, an area important to learning and memory. If stem cells are found lurking in the human spinal cord, as well as in the brain, they may be stimulated to grow after injury. In animals, stem cells in the spinal cord have been shown to divide in both white and gray matter, but these cells grow into new support cells, oligodendrocytes and astrocytes, not neurons. However, studies are beginning to suggest that scientists may be able to change the cells' mind, stimulating them to mature into neurons.

The first animal studies using stem cells in spinal cord repair are promising. In December of 1999, Dr. John McDonald and Dr. Dennis Choi at Washington University in St. Louis reported that they successfully used embryonic stem cells from a mouse to mend rat spinal cords damaged by rapid compression, or contusion. The transplanted cells matured to form neurons, oligodendrocytes, and astrocytes. Moreover, the rats regained some ability to walk. Exactly how the transplants worked is unclear: They may have secreted stimulating chemicals or rebuilt the myelin sheath around denuded nerve fibers. Follow-up research by Dr. McDonald's group demonstrated that stem cells readily generate oligodendrocytes that are capable of rapidly myelinating fibers in the damaged adult spinal cord.

Unfortunately, stem cells are not altogether innocuous. Before they are injected into people, it is imperative to eliminate the risk that they will wander to inappropriate spots in the body, giving rise to unwanted growth in the wrong place, such as skin growing in the brain, or prompting the appearance of monstrous tumors known as teratomas, composed of multiple tissues that may include hair, teeth, or bone. However, several companies and research teams have stated their intention to perform human trials with stem cells in the early twenty-first century. And in the fall of 2000, the biotech company Diacrin in Charlestown, Massachusetts, received approval from the Food and Drug Administration to perform a clinical trial using fetal neural cells, containing neural stem cells, from pigs. The trial was to involve six people with long-term spinal cord injuries and was to be conducted at Washington University in St. Louis and at Albany Medical Center in Albany, New York.

## Paving the Way

The way to stem cell transplantation into the spinal cord has been paved by fetal cell transplants, which are already performed in humans with spinal cord injury in several countries. Cells removed from the fetus are committed to a particular tissue type, such as the spinal cord, but they are still vastly more flexible and adaptable than adult cells.

As befits the tortuous history of spinal cord repair, fetal cell transplants into the mammalian spinal cord had originally been proclaimed nearly impossible. In the brain, such transplants were already shown to be successful in the 1970s: The cells matured in the damaged brain area, replacing the recipient's malfunctioning neurons, such as those affected by Parkinson's disease. The spinal cord, however, was declared "by far the most difficult region of the central nervous system to achieve successful transplantation." Nonetheless, several scientists kept trying in the hope that eventually the cells would "take." One of these scientists was Paul Reier, then at the University of Maryland. In a way, he was following up on the thread of regeneration studies created by William Windle. Reier's department head was Lloyd Guth, a disciple of Windle who had kept a spinal research program alive at the University of Maryland throughout the 1970s when hardly anybody was working on regeneration of the mammalian spinal cord. In the early 1980s, inspired by fetal cell transplants in the brain performed by Anders Björklund and colleagues at the University of Lund in Sweden, Reier focused his work on using fetal cells to repair the spinal cord.

Reier says he had originally taken up regeneration out of scientific curiosity alone but "saw the light" in terms of clinical applications of his work after being invited to address a convention of an advocacy group called the Spinal Cord Society. "For the first time I was in a room with at least 500 people in wheelchairs; I realized I'm going to get up and talk about something that's incredibly important for these people, and I knew that now I'm in this business for a different reason." Upon moving to the University of Florida in 1985, together with new colleagues, he set out to develop fetal transplants for spinally injured humans.

The University of Florida launched a pilot human trial of fetal transplants in 1997. Over the course of four years, it enrolled eight people with long-standing spinal cord injuries who needed surgery anyway because they had developed cysts at the former injury site. The cyst is an expanding cavity that sometimes forms in the spinal cord long after the initial trauma. It can catastrophically worsen the person's ability to function because it causes paralysis to creep up the spine; for example, people initially paralyzed from the waist down may lose the use of their arms. The Florida initiative followed closely on the heels of a Swedish trial, launched in 1996 by Professor Åke Seiger of the Karolinska Institute, in which three people with paralysis, who had also developed cysts, received transplants of fetal spinal cord tissue.

In both the American and the Swedish trials, preliminary results suggest the transplants may be able to prevent the spread of cyst-related damage. Fetal cells do not seem to restore body function in humans, but the trials have demonstrated that transplants into the human spinal cord are feasible and safe, thus paving the way for other cell therapies. (Fetal cells and stem cells are also being transplanted into humans in Russia, China, and several other countries, but results are difficult to evaluate because they have not been published in peer-reviewed Western scientific journals.) Fetal tissue research has shown that animal studies may lead to human trials faster than previously thought. "Fifteen years ago, I would have probably thought it would take 100 years to get to where we are now," Reier says.

## A Nose for Treatment

More cells for spinal cord repair may come from an unlikely fountain of youth, the mammalian nose. The nose is one of the rare places in the nervous system where nerve cells are renewed throughout an individual's life. This mechanism probably emerged during evolution because the sense of smell is crucial for survival; losing the nose neurons, which transmit scent information to the brain, would deprive the animal of a long-distance alert system about food and danger. As the only nerve cells in contact with the outside world, the nose neurons are easily destroyed—for example, by exposure to toxins in the air—but they are continuously replaced by new neurons, arising mainly from stem cells in the lining of the nasal cavity.

The candidates for spinal cord repair are not the nose neurons themselves but caretaker cells called olfactory ensheathing glia, which assist in the nerve renewal process. As their name suggests, these cells envelop the axons, protecting their growing tips from the hostile central nervous environment. The nose glia also nourish the axons and guide them to their targets in the olfactory bulb, a lump of brain tissue behind the bridge of the nose that relays signals to brain regions dealing with smell perception. The nose glia had been discovered in the late nineteenth century, but scientists paid no attention to these cells until the early 1980s, when their properties were first described with modern methods. Even so, the true nature of the nose glia is still somewhat mysterious. Some scientists believe they are one step away from stem cells; they may even be able to revert to their stem cell stage.

The nose cells may be able to perform the same spinal cord repair functions as Schwann cells, providing a supportive surface for regeneration, nourishing the regenerating axons, and perhaps covering them with myelin. But the nose glia have advantages over Schwann cells: They tend to move about, and they can become integrated into central nervous tissue. These navigating skills and integration ability make the nose glia particularly attractive for spinal cord regeneration; these cells may be good at coaxing spinal cord axons to grow over long distances.

Important studies on the biology of the nose glia have been carried out at several universities around the world, but attempts to use these cells in spinal cord repair were launched in a place steeped in regeneration history, the Cajal Institute in Madrid, today the major center of brain research in Spain. There, in the early 1990s, Almudena Ramón-Cueto, then a graduate student in the lab of Manuel Nieto-Sampedro, developed a method for isolating and growing adult ensheathing glia of rats in a laboratory dish and showed that these cells could stimulate regeneration of sensory fibers in the rat spinal cord. Later on, working at other institutions, Ramón-Cueto proceeded to inject the glia obtained from rat skulls into rats whose spinal cords had been completely cut. She reported in 2000 that after the treatment, formerly paralyzed rats were able to climb a vertical grid in order to collect a chocolate reward. Now working in her own laboratory at the Biomedical Institute in Valencia, she plans to expand her nose glia studies to monkeys.

Another animal study to suggest that the nose cells may bring about recovery of function after spinal cord injury was performed at the National Institute for Medical Research in London. The research was led by Geoffrey Raisman, whose electron microscope findings back in the late 1960s had helped launch the research of central nervous system plasticity. "I started working in this field 30 years ago, and only in the last 12 months have we had this evidence of nerve fiber regeneration," Raisman told the *New York Times* when his nose glia results appeared in *Science* in 1997.

More animal studies with ensheathing glia are under way in other laboratories around the world, as the cells are gaining popularity among researchers. But before the nose glia are transplanted into humans, crucial questions must be answered: Are the cells safe? Where and how exactly should they be delivered to the spinal cord? Moreover, humans rely on their sense of smell less than do animals, and scientists need to find out whether the human nose cells have the same properties as animal cells. And finally, a suitable method for obtaining human ensheathing glia must be found because opening the skull to remove the person's olfactory bulb is hardly an attractive idea.

Several methods are being explored to ensure a supply of nose glia for the repair of human spinal cords. Miami Project scientists are developing a technique to remove the person's own nose cells via a minimum opening in the skull. Scientists at the Cajal Institute in Madrid are setting up a donor bank of human ensheathing cells from cadavers, modeled on donor banks for other body organs. Researchers at the universities of Glasgow, Cambridge, and Yale have developed the first methods for growing human ensheathing glia in tissue culture and shown that these cells can remyelinate central nervous axons that have been stripped of their myelin sheath. And an alternative source, nose glia from pigs, is being developed by Alexion Pharmaceuticals, Inc., in New Haven, Connecticut.

These studies seek to bank on the evolution-bred wisdom of cells, but scientists do not always know exactly what the cells do during the repair process. To speed up the answers, many research teams attempt to cooperate with nature at the most intimate level: They are zeroing in specifically on molecules that can promote regeneration.

# 13

# Elixir of Youth for Damaged Nerves

E ver since Cajal suggested that regenerating neurons needed "special food," scientists had suspected that growth-stimulating substances in nervous tissue played a role in regeneration. The first of such nourishing chemicals made a quiet appearance at William Windle's historic regeneration conference at the National Institutes of Health in 1954. The Italian scientist Rita Levi-Montalcini presented evidence that an obscure "agent" released by fragments of mouse tumors stimulated the growth of peripheral nerves in a test tube. The substance did not yet have an official name, and no one had a clue about its enormous potential. The discovery of the "agent," soon to be called the "nerve growth factor," would later be hailed as one of the greatest biological findings of the twentieth century and would win Levi-Montalcini the Nobel

Prize. It would lead scientists to discover a huge arsenal of related molecules, now known as "growth factors."

Today potential medical uses of growth factors constitute a multibillion-dollar market and are as varied as the tissues in which the factors are found. In spinal cord injury, they can serve as a nourishing potion, saving nerve cells from death and helping them regrow cut axons. Transplants of peripheral nerves, Schwann cells, and other types of cells have been aimed, among other goals, at supplying growth factors for regeneration. Now, thanks to new molecular biology methods, the growth-stimulating molecules can also be delivered to the injury site in a pure, concentrated form. Dozens of laboratories are studying the capacity of these molecules to promote repair after spinal cord injury, and numerous growth factors are revealing their different roles in this process, including the forefather of all the growth stimulants, the nerve growth factor.

If there ever was a nerve regeneration story worthy of an epochal thriller, it is the saga of the nerve growth factor. Levi-Montalcini, now a grande dame of neuroscience and residing in Rome, was born Rita Levi in 1909 in the Italian city of Turin, into what she described as an intellectual Jewish family. She resolved never to marry and added her mother's maiden name to her surname when she began her career. Upon graduating from medical school, she hesitated between an academic and a clinical career, but her indecision was not to last long: In June 1938 Mussolini issued a manifesto that barred non-Aryan citizens from practicing both science and medicine. She proceeded to build a secret laboratory in her bedroom, equipped with primitive tools that included an incubator, a light, a microscope, and sewing needles transformed with the help of a sharpening stone into microinstruments. After Turin was heavily bombed by the Allied powers in 1942, her family moved to a small country house in the hills, where she rebuilt her laboratory in a corner of the dining room.

While riding to a small mountain village in a cattle car converted during the war into a passenger wagon, Levi-Montalcini read an article by the neurobiologist Viktor Hamburger, who had moved to the United States from Germany in 1932, one year before Hitler came to power. The 1934 article, which examined the development of the nervous system in chicks, became her "Bible and inspiration." "The wagons, which lacked seats, doors and windows, offered great panoramic views through the windowless open sides," Levi-Montalcini writes about the experience that led her to devote her research to the development of embryonic chick neurons. "I don't know how far the idyllic circumstances in which I read the article contributed to my desire to delve into this phenomenon, but in memory my decision is indissolubly bound up with that summer afternoon and the smell of hay wafting into the wagon."

Since eggs were extremely scarce, the future Nobel laureate "cycled from one hill to another begging farmers to sell me some 'for my babies.' Casually I inquired whether there were roosters in the chicken coop because, as I explained, 'fertilized eggs are more nutritious.' " (Birds in the wild, after copulation, lay fertilized eggs that give rise to fledglings. Domestic hens, in contrast, have been bred to regularly lay their eggs without participation of the roosters. These industrial, unfertilized eggs contain only the support structures for the embryo, the yolk and the white; when roosters are around, the eggs may also contain the embryo itself, a small, translucent fluid-filled balloon with a rich membrane of blood vessels. As the embryo develops, it resorbs the yolk, while using up the white for protein supply.) After removing the five-day-old embryos for research, the scientist whipped the leftover eggs into omelets.

Levi-Montalcini conducted her clandestine research with her former professor and mentor, Giuseppe Levi (not a relative), who had been forced to resign from his post at the University of Turin. At a time when the anti-Semitic campaign in Italy reached its peak, when the atmosphere was pervaded by Nazi slogans and the mountain sky was lit by the glare of fires from nearby bombings, the two scientists set out to explore one of life's most fascinating mysteries: the development of a single fertilized cell into a complex organism consisting of billions of specialized, perfectly organized cells. They focused on one specific aspect of this process: the embryonic dialogue between body tissues and nerve cells. The purpose was to find out how peripheral tissues signal embryonic nerve cells to mature into different types of neurons and to link up with appropriate organs.

Viktor Hamburger had conducted experiments in which he either nipped a budding limb from a chick embryo or grafted another limb on to see how the nerve cells adjacent to the spinal cord send out fibers into the limb. When he cut off a wing bud, the number of the sensory and other neurons that would have normally extended to that tissue was greatly reduced. Hamburger concluded that without the influence of the wing tissues, the precursor embryonic neurons failed to mature. Levi-Montalcini and her mentor repeated these experiments and obtained similar findings, but they reached a different conclusion. They suggested that embryonic nerve cells mature but die if deprived of their normal targets. This hypothesis had far-reaching consequences for understanding the nervous system's development.

Hamburger, who served as chairman of the Department of Zoology at Washington University in St. Louis, had come across the 1942 article by the two Italian scientists and was intrigued by their hypothesis, which contradicted his own. That manuscript, refused by Italian journals in view of the non-Aryan names of its authors, was published in the Belgian *Archives de Biologie*. A year after the war had ended, in the spring of 1946, Hamburger invited Levi-Montalcini to come to St. Louis in order to reinvestigate the problem.

Levi-Montalcini's research at Washington University took a new direction after Professor Hamburger brought to her attention a bold and imaginative experiment conducted by one of his former students. The experiment consisted of grafting pieces of a mouse cancerous tumor onto the membrane that envelops the chick embryo. (Because embryos do not possess the immunizing properties of adult organisms, they do not reject the tumor.) The study revealed that embryonic nerve cells spread like wildfire into the mouse sarcoma, in larger numbers than on the side where the tumor was not implanted. Levi-Montalcini repeated the experiment and found that the same thing happened even when the tumors were not in direct contact with the embryos. She had a brilliant, prophetic

*Target cells secrete NGF*

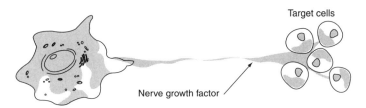

*Axon grows toward target cells in response*

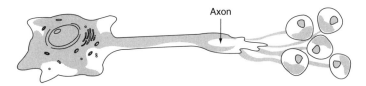

*Axon forms synapses connecting them with the target cells*

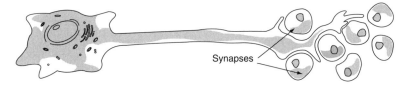

Nerve growth factor (NGF) plays a crucial role in the organization of the developing nervous system. By secreting NGF, target cells can determine how many neurons survive, grow toward them, and establish connections with the appropriate cells. (Redrawn with permission from *Scientific American.*)

insight: that the tumor released a growth-stimulating chemical of an unknown nature, soluble in tissue fluids.

The turning point in the investigation came in 1952 when Levi-Montalcini, carrying in her handbag two mice transplanted with sarcoma cells, traveled to Brazil in order to continue the experiments using a faster and simpler method. The method was tissue culture, or growing pieces of tissue in laboratory glassware. In the early 1950s tissue culture had yet to become the universal biological tool it is today and Washington University did not have the appropriate facilities, which were available at a medical school in Rio de Janeiro, where a friend of Levi-Montalcini's, a refugee from Nazi Europe, had set up a laboratory. Levi-Montalcini found that nerve cells cultured in the test tube in the proximity of tumor fragments sent out fibers that spread out "like the rays of the sun," forming a halo with maximal density on the side facing the tumor. The approach made it possible to quickly screen tissues to determine whether they were sources of the growth-promoting activity.

The way the nerve growth factor was eventually identified was a case of incredible scientific serendipity. A young biochemist named Stanley Cohen, who worked with Levi-Montalcini on the project at Washington University, isolated the portion of mouse tumors responsible for the growth-promoting activity and wanted to break down the isolated substance further with enzymes, but because enzymes were still extremely difficult to obtain, he was advised to use snake venom instead. Amazingly, even in minute amounts, the venom turned out to cause the nerve cells to send out fibers in a halo of tremendous density. Bizarre as this may sound, snake venom revealed itself as a rich source of nerve growth factor, at a concentration estimated as a thousand times higher than that in the mouse sarcomas. (Nearly 50 years later, it is still unknown what the nerve growth factor is doing in snake venom. According to one hypothesis, the molecule, which specializes in interacting with the nervous tissue, may simply serve as a carrier that helps the snake's toxins latch onto the nerves of the bitten victim.)

The third potent source of the nerve growth factor, after tumor cells and snake venom, proved to be no less strange; it was the mouse salivary gland, related in function to the snake venom gland (in addition to digestive enzymes, mouse saliva contains toxins instilled into the victim bitten by the mouse). Cohen then produced a key piece of evidence that the growth factor is required during normal development: When he injected antibodies against the nerve growth factor into newborn mice, the rodents failed to develop certain neurons.

# Preventing Cell Suicide

At the beginning, the nerve growth factor, or NGF, received a cool welcome from the scientific community; its sources were puzzling, and its scope of action was unclear. The first method for obtaining NGF in large quantities, which made it available for more extensive research, was developed in the 1960s at Stanford University. But in 1977 *Science News* still ran the article: "NGF May Hold the Key— But to What?" It was only in the early 1980s that NGF was shown to be an extremely important molecule in many processes. It was found to be secreted in small amounts by a wide variety of normal and cancer cells, and it was shown to play a role in the central nervous system, not only in peripheral nerves. Moreover, NGF gradually revealed itself as one of the body's many growth factors that are active throughout adulthood and not only during the development of the fetus.

The second member of the NGF family, purified in 1982, was found in the brain of adult pigs and called the "brain-derived neurotrophic factor," or BDNF. (The word *trophic* is derived from ancient Greek for "nourishment"; growth factors in the nervous system are referred to as trophic or neurotrophic factors.) In the following few years, scientists proceeded to search the central nervous system for more growth factors and found so many of these scarce molecules that they were simply numbered: neurotrophins 3, 4/5, 6, and 7 (neurotrophins 4 and 5 were purified by two separate research teams and later turned out to be the same molecule, hence the name, neurotrophin 4/5).

Growth factors are proteins made by the body in minute amounts, but even a few billionths of a gram suffice to produce a powerful effect on a cell. The factors act through tiny molecules on the cell surface, called receptors. A growth factor slips into the receptor molecule, much like a space module docking with the mother spacecraft; the "docking" sets off a chain of biochemical events inside the cell, and these events ultimately activate some of the genes in the nucleus. Depending on which genes are switched on, the cell may begin to divide or grow.

The role of growth factors in the nervous system is particularly dramatic during the development of the fetus. In vertebrates, including humans, the nervous system begins as a relatively straight tube that grows rapidly as its cells migrate to their new homes, giving rise to the brain, spinal cord, and other nerve tissues. (In the developing human embryo, brain cells are produced at a staggering rate of 250,000 per minute.) The fast-growing tissue is a scene of a mass suicide: About half of the fetal neurons die. This happens because body tissues release a limited supply of growth factor molecules, which signal to the neuron: "Stay alive, you are in the right place." Only the neurons that receive this chemical call survive; the unlucky ones that fail to hook up with a growth factor activate a built-in suicide program.

What at first glance seems like a terrible waste of cells is in fact a highly efficient way of putting the nervous system together. A newborn neuron is oblivious of its fate until it successfully competes for a growth factor. In this manner, each tissue, by releasing growth factors and other chemicals, dictates how many nervous fibers it needs and where; the tissue does not determine how many neurons are generated, but it does determine how many survive. This explains, for example, how the human spinal cord ends up sending more axons to body areas with more muscles and fewer axons to areas with fewer muscles. The details of this mechanism continue to be investigated, but its essence had been captured by Levi-Montalcini in her wartime hypothesis, the one that won her the invitation to St. Louis in the 1940s.

In the adult nervous system, the name "growth factors" covers a tiny fraction of what these molecules do. While many are shut off before or shortly after birth, others perform maintenance functions throughout adulthood and keep nerve cells alive. They are believed to play a role in learning and memory, and in rat experiments, NGF has even improved the performance of memory-impaired animals. Some neurotrophins also affect tissues outside the nervous system.

In 1986, Professors Levi-Montalcini and Cohen were awarded the Nobel Prize in physiology or medicine. Levi-Montalcini's discovery of NGF was called "a fascinating example of how a skilled observer can create a concept out of apparent chaos." The Nobel committee also noted that the two laureates "have created a scientific school with an increasing number of followers." (Many neuroscientists believed Viktor Hamburger deserved to share the Prize for his early contributions to the NGF discovery. The Nobel committee's decision to exclude him, which stirs up discussion even today, reportedly put an end to the friendship and close collaboration that had united Levi-Montalcini and Hamburger for 40 years.)

What had started as a single experiment once described as "peripheral in every sense of the word," turned into a booming field with hundreds of biotechnology companies and noncommercial research teams studying, developing, or producing growth factors around the world. Growth-factor research exploded in the late 1980s and early 1990s thanks to recombinant DNA technology, which allowed researchers to isolate the genes for growth factors and manufacture them in large quantities.

The nervous system's neurotrophins form but one subset of growth factors; many other families are found in different tissues. Their current and potential uses include treating anemia, speeding wound healing, repairing broken bones, or growing skin grafts for burn patients. Because growth factors play a role in cancer—they can transform healthy cells into cancerous ones—it may be possible to stop the spread of malignant disease by blocking the stimulators of aberrant growth.

## Food for Growth

Researchers have high hopes for growth factors in spinal cord repair. One major problem in cord trauma is the actual death of cells. Neurons, in particular, often shrivel or die when their axons are cut, probably because without the healthy axons they lose their supply of growth factors. It is as if, without the factors, a cell body loses its reason to live, and once the cell body is gone, there is nothing left to regenerate. Therefore an important line of research focuses on using growth factors to keep nerve cell bodies alive or to revive them in the injured spinal cord.

Scientists also increasingly incorporate growth factors into their attempts to regenerate spinal cord fibers. However, attractive as this may sound, simply flooding the spinal cord with growth factors is not a good idea. The molecules are far from innocuous: They may, for example, stimulate unwanted growth, leading to pain. In fact, in several cases human trials with growth factors in neurological disorders had to be discontinued because of adverse effects: Pain was a serious side effect when researchers first tried pumping growth factors into the brains of patients with Alzheimer's disease in the 1990s, while growth-factor trials of patients with Lou Gehrig's disease had to be discontinued after they caused weight loss and general debilitation. Thus, a way must be found to deliver the stimulating molecules to the spinal cord in high doses but gradually and only to the narrow region where they are needed. In the 1996 Karolinska study, a growth factor was delivered to the rat spinal cord using Dr. Henrich Cheng's ingenious "dumpling" method, which is, unfortunately, difficult to apply. A more viable alternative may be a sophisticated molecular biology approach: gene therapy.

When the first gene therapy study in the spinal cord was reported in 1994, the wartime-thriller growth factor, NGF, was the star. A team led by physician-researcher Mark Tuszynski, in collaboration with the lab of Dr. Fred Gage, Tuszynski's former Ph.D. adviser, had been developing a gene therapy for Alzheimer's disease at the University of California, San Diego. Seeking to avoid the side effects encountered in the earlier Alzheimer's trials, the researchers devised a way to deliver the growth factor in a controlled, targeted manner, rather than freely pumping it into brain tissue. They put the NGF genes into skin cells called fibroblasts, which divide aggressively, and transplanted these cells into the brains of laboratory animals. Some of the cells were left over from the study, and Tuszynski, who had long wanted to start working on spinal cord regeneration, used the opportunity to design a spinal cord experiment: "I said, let's put these cells into the injured spinal cord of a rat and see if they do anything. We expected to see maybe a little bit of growth of spinal cord fibers into a transplant of these genetically modified cells, and lo and behold, instead we saw a huge amount of growth."

Encouraged by the magnitude of NGF's effect, Dr. Tuszynski's team designed another spinal cord experiment; they modified cells to include the gene for a growth factor believed to be particularly important for the spinal cord, called neurotrophin NT-3. Paralyzed rats treated with the genetically engineered cells partially recovered the use of their limbs. This was the first time gene therapy produced such an effect in paralyzed rats and, no less important, the recovery took place thanks to the growth factor alone. These and other results suggest that growth factors may help achieve a goal that has so far proved elusive in most regeneration studies, producing massive regrowth of fibers in the spinal cord. This research, however, still raises questions to sort out: Different types of neurons respond to different growth factors, and scientists are trying to elucidate the neuronal preferences. By giving several growth factors in combination, scientists may be able to stimulate regrowth in numerous tracts of the spinal cord simultaneously.

Growth factors, in a way, put a nerve cell through a rejuvenation program. A variety of genes are active producing growth factors during fetal development, but once the cell stops growing, these genes shut down. They switch on again after injury in peripheral nerves but not in the central nervous system. Growth factors captured in the laboratory, therefore, can serve as an "elixir of youth" for the injured spinal cord, compensating for the silence of growth-stimulating genes in adulthood. But how can the growth factors stimulate regeneration in the presence of all the inhibitors, such as those present in myelin, that create a barrier for regrowth?

One intriguing possibility is that under certain conditions, growth factors may allow the regenerating axon to ignore or perhaps even neutralize the inhibitors. Dr. Barbara Bregman, of Georgetown University, who has studied the role of growth factors in spinal cord repair for many years, discovered that spinal cord neurons exposed to certain growth factors were capable of growing through myelin-containing white matter. And in a study of the rat brain conducted by Professor Martin Berry of King's College, London, and his colleagues, growth factors appeared to neutralize the chemical scar that forms after nerve injury. According to a recent hypothesis, each neuron contains a central program, a series of genetic "master switches" that may sometimes allow the neuron to ignore obstacles and move ahead like a bulldozer plowing its way through rough terrain. Says Lisa McKerracher of the University of Montreal, one of the strongest proponents of this idea: "A neuron encounters various positive and negative signals in its 'environment' and needs to make the decision: Do I grow or not? We may be able to affect this decision by identifying the 'master switches' involved." Several such switches and their molecular "helpers" have already been discovered by different laboratories and at least three—called Rho, cAMP, and inosine—are showing promise in spinal cord repair studies in laboratory animals.

Initial experience with giving growth factors to people with nerve diseases has shown that the substances must be used with caution. But if these molecules make good on their promise, they may prove a powerful force in putting the spinal cord back together again after injury. Meanwhile, the same virtuosity with molecules that scientists brought to growth factors has created the possibility of outfitting the injured spinal cord with other, entirely new means of repair.

# 14

# Wired for Regeneration

Ahot new field of neuroscience seeks to explain how nerve networks come into being and is opening up the intriguing possibility of healing the injured spinal cord with the same tools that nature used to wire it up in the fetus. The study of axonal guidance, as this area of research is known, took off in a major way only in the mid-1990s, but it may very well hold the final key to the future of spinal cord regeneration. Studying axonal guidance may allow scientists to simultaneously go after two major goals that until recently were usually pursued in separation: to stimulate the growth of axons and to eliminate the mechanisms inhibiting this growth.

The fetal nervous system is a busy place. It swarms with billions of hairy nerve fiber tips that crawl about until they hook up into elaborate neuronal circuits. The result of this wiring is a network that, in humans, holds the distinction of being the most complex object in the known universe. The human brain alone has more than 100 billion nerve cells, about 10 times more than the number of stars in our galaxy. Some simpler neurons have "only" 100 connections with other neurons, but many have 1,000 to 10,000, or even more. How are circuits of such mind-numbing complexity formed? Apparently, growing axons are guided to their new homes by chemicals that have the effect of far-reaching "odors." The growing axons, like living beings with a strong preference for particular odors, appear to be attracted by some chemicals and repelled by others.

Once the development of the nervous system is completed, the nerve fibers stop their crawling about. Scientists suspect that the same mechanisms that "freeze" the crawling remain present throughout adulthood and may inhibit effective regeneration after injury. When the Swiss scientist Martin Schwab zeroed in on the first growth-inhibiting mechanism in the adult spinal cord in the mid-1980s, he was hoping to find one regeneration-blocking molecule. He identified the protective myelin sheath as a "fixative" blocking nerve growth in the adult spinal cord and spent 15 years scanning myelin for the inhibiting molecule he called Nogo. Now it appears that the brain and spinal cord are also strewn with numerous molecules that helped wire up the system in the fetus but "froze" it after the development was completed. To produce regeneration, scientists would have to "unfreeze" the spinal cord, unleashing the guiding chemicals and inviting the cord to be "born again." After all, during regeneration the axons need to repeat what they once did in the developing nervous system: grow and arrange themselves into functioning circuits.

The idea that chemicals control the wiring up of the nervous system goes back to Santiago Ramón y Cajal. In 1892, relying entirely on observation and intuition, Cajal postulated that chemicals of an unknown nature guide growing axons in the embryo—and consequently, during regeneration. The tools to identify the guidance molecules were unavailable at the time, and a competing theory took over in the 1920s and 1930s. It stated that during development growing axons form random connections that are later transformed by experience and that in the course of this process, only the appropriate connections survive. Scientists were claiming that growing nerve fibers found their way to appropriate targets by following physical pathways, or grooves, in the developing tissue.

In the early 1940s, a future Nobel laureate, Roger Sperry, then at the University of Chicago, revived the research into chemical guidance of growing and regenerating axons. Sperry would later be called "the rebellious graduate student" because he began these studies while still working on his doctorate and challenged the views of his teachers, among them some of the strongest proponents of the physical guidance idea. He devoted more than a decade to nerve regeneration studies in fish, frogs, and salamanders and showed that growing nerve fibers did not follow a fixed mechanical path but appeared to carry chemical labels that enabled them to sort each other out.

In one of his most famous regeneration studies, Sperry showed that even when regenerating fibers are artificially deflected from their course, they adopt tortuous routes and find ways to grow back to their normal targets. In one experiment, he cut the optic nerve of a frog, which, unlike that of mammals, can regenerate and restore normal vision. He then rotated the eye 180 degrees, waited for the nerve to regenerate, and submitted the frog's vision to a simple test. When presented with a bug, the frog flicks its long tongue to snap its favorite delicacy; if it can see well, it won't miss. Sperry found that the regenerated optic fibers from the frog's rotated eye followed a roundabout route to their original targets in the brain. For example, the axons from what had originally been the bottom side of the retina restored connections with the part of the brain normally responsible for receiving signals from the bottom of the retina. As a result, the frog now saw the world upside down and backwards; the floor appeared to be above its head and the ceiling below, and on the vision test the frog acted accordingly: It erroneously flicked its tongue downward to catch the bug above its head and upward to catch the bug below. The preprogrammed wiring had been restored with ruinous consequences for the frog's ability to feed itself. These results suggested that the axons did not follow a predetermined mechanical path but searched out the appropriate target, apparently guided by chemical cues.

Such biochemical recognition is now known to lie at the basis of connections formed by neurons in development and in regeneration. In fact, some of Sperry's devotees believe his contributions to this field were worthy of a Nobel Prize, but he won the award in 1981 for a different achievement. (That research, conducted by Sperry later in his career at the California Institute of Technology, focused on the so-called split brain and showed that although the brain's two hemispheres work closely together, consciousness and awareness seem to exist separately and independently in each hemisphere.) Sperry's research did not immediately help revive Cajal's ideas, and chemical attraction remained extremely unpopular for several decades despite accumulating evidence in its support. It was only in the early 1990s that the search for guidance chemicals began in earnest.

One of the studies was initiated in 1990 by Marc Tessier-Lavigne, a neurobiologist in his early 30s who had just set up his own laboratory at the University of California in San Francisco. The Canadian-born Tessier-Lavigne had started out studying mathematics, then had become captivated by the circuitry of the brain and done postdoctoral studies in neuroscience at Columbia University. When he announced his intention to look for the guidance molecule, some seasoned researchers were skeptical. The chemical attraction idea was still controversial, and finding one of the guiding molecules, if they existed, seemed like a long shot.

Tessier-Lavigne decided to focus on the favorite spot for axonal guidance studies, the midline of the embryo. This is an area where growing axons behave in a dramatic and stereotypical manner. On a particular day of development, some of the axons on each side of the embryo turn their growth cones toward the midline, as if responding to an irresistible appeal, and head in its direction until they cross over to the other side. The result is a crisscrossing pattern of axons in the middle of the brain and spinal cord. In the brain, some axons from the left hemisphere cross over to the right, while some of those from the right cross over to the left. This is why in stroke, for example, people with damage to one side of the brain have impairment on the opposite side of the body. No one knows why the nervous system is organized in this manner, but hypotheses abound. One plausible explanation is that the crisscrossing of axons enables the right and left halves of the nervous system to communicate. A simpler, somewhat unromantic, theory states that the crisscrossing keeps the two halves from falling apart.

Tessier-Lavigne wanted to identify the guidance signal that attracts axons to the midline in vertebrates and decided to conduct his research on chick embryos. In case the guiding molecule was scarce, he reasoned, it would be easier to obtain large numbers of embryonic brains from chicks than from rats or mice. This turned out to be a wise decision because eventually it would take 25,000 brains to identify the guidance signal. Once a week for several months, a truck carrying cartons with 1,000 fertilized eggs would climb up the scenic streets of San Francisco and pull up behind the gray, 14-floor university building housing Tessier-Lavigne's laboratory. All the eggs contained embryos that were 10 days old, an age at which embryonic brain tissue was found to attract axons in a laboratory dish. The research team included two postdoctoral fellows and three graduate students, but when an egg shipment arrived, everybody—other students, friends of students, family members—was recruited to help. On what was called "the Bastille day" of the week, some 10 people would line up to crack the shells of 1,000 eggs, extract the embryos, and dissect out their brains. The decapitation created a formidable mess after which the laboratory required a thorough cleaning, but it produced half a beaker containing 1,000 embryonic brains for the study.

Within two years—an extraordinarily short period considering the novelty of the field and the enormity of the challenge—the team purified one protein, out of some 50,000, that was making the attraction happen in the chick embryonic nervous systems. "They say luck favors a prepared mind, and that may have been true in our case, but it would be silly to deny there is some luck involved in research—and we were obviously incredibly lucky," says Tito

Serafini, former postdoctoral fellow in Tessier-Lavigne's team. The scientists then cloned the gene that carries the code for the protein, and called it *netrin*, after the Sanskrit root *netr*, meaning "one who guides." The *netrin* report appeared in the journal *Cell* in August 1994, roughly 100 years after Cajal had formulated the chemical guidance idea.

## The Likeness of Being

The scientific community was spellbound by the discovery of a molecule that attracted growing axons in the embryo. But the true bombshell was dropped when the scientists realized that *netrin* was strikingly similar to an axonal guidance gene identified two years earlier in a microscopic worm. The similarity suggested that some of the same mechanisms are used to put together the nervous system in worms and in vertebrates!

Many early studies of axonal guidance, from the 1960s through the 1980s, had been conducted on invertebrate creatures, such as insects and worms, but that research seemed hardly applicable to vertebrates, including humans. One object of study had been the microscopic worm *Caenorhabditis elegans* consisting of fewer than 1,000 cells. Its nervous system, which consists of only some 300 neurons, provided a convenient research model, yet it seemed hardly a match for the awesome neural arrangements of more complex organisms. Although some of the scientists conducting this research believed their work would eventually be relevant for all organisms, few colleagues took them seriously. When Edward Hedgecock of Johns Hopkins University, who had performed some of the first axonal guidance studies in worms, once expressed the idea in writing, a colleague who reviewed his paper suggested removing the passage because, he said, the parallel between worms and vertebrates was too far-out. Hedgecock's collaborator, Joe Culotti of the University of Toronto, whose lab welcomes visitors with the sign "We have worms," recalls: "When we talked about it people would sort of chuckle, 'Oh, right, sure,' but we were quite serious. We always thought there would be similar genes that would play similar roles in worms and in the vertebrate spinal cord. What we didn't imagine was that they'd be discovered so soon in vertebrates. We thought we had more years to play around with the genetics of the worm and just live in our own little corner of the universe, building the story up."

The worm gene found to be similar to *netrin* belonged to a group of genes named *uncoordinated*—or *unc* for short—because when these genes were missing or did not function properly, the worm's axons became messed up and the animal did not wiggle properly. Teams of scientists headed by Hedgecock and Culotti had established that several of the *unc* genes could account for axonal guidance problems in the *C. elegans* worm. When the *netrin* report came out, findings from the *netrin* and *unc* studies were found to complement each other in many ways, and together they created a fuller picture of axonal guidance. The report prompted a spate of other comparisons among guidance molecules that were beginning to be discovered in different organisms, and numerous similarities emerged. The same genes, with minor variations, were found to play similar roles in the nervous system's wiring in worms, insects, birds, and mammals.

It was a humbling but also a fantastic discovery. Corey Goodman of the University of California at Berkeley, a veteran investigator of axonal guidance whose own work had dealt with fruit flies, summed up the discovery's implications in the title of an editorial as the "likeness of being." Worms seem light-years away from vertebrates—particularly the species engaged in neuroscience experiments—yet some of the tools used to guide developing axons have been preserved in the course of evolution and are strikingly similar in these different creatures. The finding was all the more remarkable considering that worms and vertebrates diverged on the evolutionary tree some 600 million years ago. Scientists studying different organisms could now trade insights and make rapid progress in deciphering how the nervous system is formed.

As befits a mechanism perfected over 600 million years, axonal guidance is a smoothly scripted affair. Axons travel a long distance, sometimes more than a thousand times greater than the diameter of their cell bodies, before settling into their assigned spots. They negotiate this challenging journey by breaking it up into short segments, each perhaps a fraction of a millimeter long. At the end of each segment, the growth cone appears to pause, like a traveler at a crossroads, making navigating decisions. What helps it choose its course is the presence of guidance chemicals. Each growth cone is equipped with numerous receptors that allow it to sense attractive and repulsive guidance molecules released by distant tissues. Scientists have found support for Cajal's idea that the chemical guidance is "graded": If an appealing "odor" is diffused throughout the tissue, axons are attracted toward its greatest concentration. If, for example, only an attractive chemical is present in the vicinity, the axon will migrate all the way to the highest concentration of this chemical. If it senses both attractive and repellent chemicals, it will migrate to the point where attraction precisely balances out the repulsion. In addition, the axon is sensitive to the underlying surface, which is studded with molecules that also attract or repel the axon. This surface needs to be sticky, or adhesive, to the proper degree. The fibers get bogged down on a material that is too sticky, but neither can they grow over a smooth surface like glass, which provides them with nothing to grab onto. Because the molecules in the surface material are fixed in place, they do not operate over a long distance, only upon contact or when the axon is within a short range. "Thus, an individual axon might be 'pushed' from behind by a chemorepellent, 'pulled' from afar by a chemoattractant, and 'hemmed in' by attractive and repulsive local cues. Push, pull and hem: these forces appear to act together to ensure accurate guidance," Tessier-Lavigne and Goodman wrote in a 1996 review.

Where is the blueprint for this orchestrated deployment? Its main lines are encoded in the genes, but one will never find an entire map of the nervous system in the genetic code. This is because the system's blueprint is not a static map with all details spelled out in advance. Rather, it is more like an interactive program that unfolds gradually, constantly responding to cues from the environment and making sure different genes are switched on and off at the right times.

Each axon does not have a specific gene-encoded "address" telling it in advance where to go. Instead, at different times, the "program" prompts the axon to display different receptors on its surface, altering its response to the surroundings. The changeover in receptors may not be necessarily programmed in advance; it can occur in response to signals arriving from the environment. This means that in identical twins, who have the same genes, the final arrangement of the nerve circuits will be at least somewhat different. The midline and other tissues releasing guidance signals also appear to change their personality at different stages of development by displaying different cues and affecting axonal navigating decisions. Thus, as the axons approach the midline, they find it irresistibly alluring, apparently because they are drawn by netrins. But once they have crossed the midline, the axons turn sharply and run along the length of the body on the other side or toward a different destination, and they never cross back. Not only do they no longer find the netrins attractive, they now find them repulsive.

Scientists are only beginning to uncover the details of this intricate topography. No one knows how many guidance molecules exist, but several dozen have already been found and others keep being discovered. Their names often reflect their function. One family is called *semaphorins*, after the word *semaphore,* meaning a signaling system. A gene *robo,* for *roundabout,* the British term for a traffic circle, is so called because when mutated, it causes fruit fly axons to cross the midline back and forth. Mutation in another gene—*commissureless,* or *comm* for short—leads to the opposite problem, a lack of axonal crossovers, or commissures. Examples of other evocative gene names include *beaten path, sidestep, frazzled, collapsin,* and *connectin.*

## Turning Repulsion Into Attraction

Axons have no problem following the attractive and repulsive cues in the embryo, but by adulthood something happens to the environment in the central nervous system, making it hostile to growth. Scientists hypothesize that the "go" signs that guided axons to their places in the fetus turn into "stop" signs in the adult organism. To produce regeneration, the guiding signs would have to be switched from "stop" back to "go." This idea is based on a fascinating property of all guidance molecules: They can be attractive or repellent at different times, or attractive for some growth cones but repulsive for others, which suggests that their guiding properties can be manipulated.

Scientists have proposed that the molecular "stop-go" switch may be the same for all guidance molecules. No one has yet proved this hypothesis, but in several studies a manipulation of guiding properties has already been accomplished. In one series of experiments, a team led by Dr. Mu-ming Poo from the University of California at San Diego, in collaboration with Tessier-Lavigne's laboratory, achieved a feat that sounds like an episode from a romantic novel. The scientists managed to transform repulsion into attraction. Using a molecular switch they identified, they altered the effect of two guidance molecules on the tips of growing nerve fibers: Instead of repelling the growing fibers, the molecules started to attract the fibers in a laboratory dish.

Scientists are trying to identify the key targets for such manipulation. Netrins, first identified in worms and chick embryos but later found to be present in fruit flies and in humans, are among the potential candidates. Timothy Kennedy, former postdoctoral fellow from Tessier-Lavigne's San Francisco netrin team, who now runs his own laboratory at McGill University's Montreal Neurological Institute, has shown with his colleagues that netrins are present in large amounts in the adult spinal cord of rats and that their amounts change after injury. During development, netrins have been shown to attract the growth of some axons and repel the growth of others, but Kennedy believes that their role in the adult nervous system may be to block nerve fiber growth. By cranking up the attractive properties of netrins, he suggests, it may be possible to encourage spinal cord axons to regenerate in adults.

Producing massive growth in the spinal cord, however, will only be useful if the growing fibers are properly rewired. The chances for such rewiring would increase if the fetal road map for growing axons were preserved throughout an organism's life. Several studies have suggested that when fetal neurons or stem cells are transplanted into the injured adult brains or spinal cords of rats, they deploy themselves just as they would have in an embryonic nervous system. It is possible that the transplanted cells and their fibers rely on navigating instructions that are still present in the adult system and can follow these instructions in order to stay on the right track, at least when in close proximity to their targets. This navigating savvy, in turn, increases the chances that the new nerve fibers may be capable of making the right connections for restoring function after injury. "I think we will be able to help rewire parts of the brain or spinal cord following injury," says Tessier-Lavigne. "Our best hope is that enough guidance information remains in the adult nervous system to guide regenerating fibers back without us having to assist in this process. Our major task will be to get lots of regrowth, and the nervous system can then make sure it's the right kind of regrowth, so that it can be translated into an effective therapy."

# 15

# The Smart Spinal Cord

I f researchers tried to restore the spinal cord to precisely the way it was before the injury, the task could be declared hopeless in advance. Replacing all the dead cells, regrowing all the cut fibers, and reconnecting all the circuitry is hardly a feasible proposition; the human spinal cord contains millions of fibers, and Nature forgot to color-code them. Fortunately, such ambitious undertakings may be unnecessary. While only recently the spinal cord was relegated to a Cinderella status in the central nervous system, as an appendage entrusted with the plain chore of passing signals between the body and the brain and producing simple reflexes, now scientists believe they are dealing with a smart and sophisticated system that may be able to adjust to the postinjury situation. The mammalian cord, it turns out, contains much of the circuitry needed for walking and can perform complex tasks with minimal input from the brain. What's more, it is capable of fulfilling such intelligent duties as processing information or even learning, which suggests that after injury neural circuits in the cord can be "taught."

The new appreciation of spinal cord intelligence may ultimately transform the entire attitude toward recovery from spinal cord injury. Physicians have long recognized that, in stroke, many patients recover when healthy parts of the brain take over from damaged ones, but spinal cord injury was traditionally viewed as irreversible. Rehabilitation meant teaching people to adjust to a wheelchair, and physical therapy was limited to strengthening healthy muscles. This attitude is now changing. For one thing, recovery stories have become more common; a larger proportion of cord injuries today are partial, possibly the result of seat belt use and improved emergency care, and people with such injuries sometimes recover remarkably. Moreover, in an unusual convergence of basic research with rehabilitation, the new science of spinal cord circuitry has led to an experimental therapy that has already gotten several people with new or long-term spinal cord injuries out of wheelchairs. Among them are people who had been told by their doctors they would never walk again.

The therapy, which consists of helping the body to relearn how to stand and walk while stepping on a treadmill, at present is available mainly to people with partial injuries, who have at least some fibers connecting the brain with the spinal cord. It is unknown whether people with the so-called complete injuries, who have neither feeling nor movement below the injury, can relearn walking, which requires at least some voluntary control of the legs. However, the hope is that regeneration therapies, if they work in humans, will turn many "complete" injuries into "incomplete" or partial ones. Then, with the help of a few regenerated fibers, the spinal cord may be taught to restore to the injured person at least a limited ability to walk or move about.

## Performing "Brainy" Tasks

Only recently have scientists realized that the human spinal cord deserves much more credit for walking than it used to be given. In other species, the circuitry for walking had long been recognized to reside mainly in the spinal cord. A headless chicken running is one gruesome example of this circuitry at work, and in rats, "spinal walking" is a well-known phenomenon that makes evaluation of regeneration therapies notoriously difficult. In humans, however, the brain is so highly developed that scientists assumed it had taken over even the basic behaviors. But now it turns out that the human brain doesn't have to bother with the multitude of commands required for the execution of each step: walking or, for that matter, running or dancing, involves dozens or even hundreds of muscles, which must contract or relax at just the right times, in a coordinated manner, for a step to take place. Like a good manager, the human brain issues one command—say, "Dance!"—and delegates the supervision of the details to the "lower-echelon" nerve networks in the spinal cord. This is probably why we can walk or dance at the same time we are thinking, talking, or for that matter, chewing gum. This general setup appears to be similar in all vertebrates, from fish to primates, but how does the spinal cord regulate walking?

Scientists think the answer may be in something they call "central pattern generators" and sometimes refer to as "motor programs." Animal studies provide intriguing clues suggesting that the cord contains elaborate networks that can govern the complex movements of the trunk and limbs. In mammals, evidence for the existence of such generators is still indirect, but simpler animals reveal what the spinal cord circuitry can accomplish on its own and how. Professor Sten Grillner and colleagues at Sweden's Karolinska Institute have mapped out the central pattern generator in a prototype kind of vertebrate, the lamprey, a 450-million-year-old eel-like jawless fish that sometimes ends up as a pickled hors d'oeuvre served with vodka in Russia and in Scandinavian countries. Grillner had started out probing the existence of the central pattern generator in cats but switched to the lamprey, which has only about 500 nerve cells in each of its spinal cord segments, as a shortcut to understanding the vastly more complex mammalian cord. In more than 20 years of research, he and colleagues have mapped out networks of lamprey neurons wired together to perform specific tasks, which consist of contracting and extending the muscles that produce the animal's wavelike swimming. The researchers found that when a signal arrives from the brain, "local teams" of neurons in the spinal cord are capable of processing sensory input, adjusting their own behavior, and coordinating the muscle movements. Having figured out the lamprey's central pattern generator down to individual cells, the scientists are now focusing on pinning down the molecular signals that set the generator in motion. While the walking circuitry in the mammalian spinal cord is still unexplained, Grillner's reasoning goes as follows: If extraterrestrial researchers wanted to figure out the workings of a car, they would fare best with a Model T Ford, which is much simpler than a twenty-first-century Ferrari but has all the essential components of an automobile. "Evolution rarely throws out a good design," Grillner wrote in *Scientific American.* "It would be most surprising to discover that there were few similarities between lampreys and humans in the organization of control systems for locomotion. . . . A turbocharged Ferrari is, after all, just another kind of car."

After injury, the spinal cord can do its job while being connected to the brain by a fraction of the normal number of nerve fibers. In humans, it is known that 50 percent of the fibers suffice to preserve normal function in the lower body because people with the so-called Brown-Séquard syndrome, which involves damage to half of the spinal cord, can recover from paralysis almost fully. Animal experiments suggest that this percentage can be much lower. In the 1950s, William Windle had shown that cats retained some walking ability when fewer than 10 percent of their spinal cord fibers were intact. Windle's findings were confirmed in other animal studies, and there is hope that in humans, too, the spinal cord can adapt to functioning even with a drastically reduced input from the brain.

Commands from the brain can reach the spinal cord via several tracts, bundles of nerve fibers that start in different areas of the brain and end in particular segments of the cord. The exact route by which the commands travel is unknown, but the mechanisms underlying movement reveal a tremendous amount of redundancy, possibly because movement has been vital in evolutionary terms:

An organism can survive only if it can chase food and escape from danger. Therefore, the brain and spinal cord can sustain a great deal of damage before the ability to move completely disappears. In a healthy person, the redundancy translates into great versatility. While having a cast on one leg, wearing high heels, or imitating Charlie Chaplin, people can walk in a variety of different ways. After injury, redundancy becomes a reservoir of backup mechanisms, and some tracts in the spinal cord probably take over from others.

This adaptability suggests that the spinal cord may be no less malleable and changeable than the brain. The cord, in a way, is catching up with the plasticity revolution, which a couple of decades earlier taught scientists to stop seeing the brain as rigid and hardwired. Thus, they now envision pattern generators in the cord not as sets of hardwired networks, but as malleable, adaptable systems. Moreover, the cord is not as separate from the brain's lofty life as previously thought. While the brain may hold the prerogative over thoughts and emotions, the spinal cord is capable of performing its own share of "brainy" tasks. Learning, for instance, traditionally thought to take place in specialized brain centers and stored, in the form of memory, in designated areas, is now believed to take place all over the nervous system, including in the spinal cord. "Learning occurs in many locations, depending on what is learned," says Jonathan R. Wolpaw, of the New York State Department of Health's Wadsworth Center, a prominent champion of spinal cord plasticity.

Dr. Wolpaw has described his findings in the area of spinal cord learning as a "continuing series of surprises." When he and colleagues trained rats and monkeys to change the size of the knee-jerk reflex in exchange for receiving a reward, the training produced anatomical and physiological changes in the spinal cord. Results of the training persisted after the brains of experimental animals were disconnected from their spinal cords, suggesting that the memory was stored in the cord and that at least some of the learning had taken place there. Moreover, complex changes in the way neurons communicate occurred throughout the animals' nervous system— probably because the system needs to readjust in order to preserve old behaviors while learning a new task, simple as this task may be. Motor learning in humans also appears to produce widespread changes in nerve circuitry. For example, University of Copenhagen researchers found that the knee-jerk reflex in dancers of the Royal Danish Ballet was smaller than in nondancers. It is possible that years of dance training reduced the reflex to make room for the precise movement control required for ballet. The need for such a reorganization of the nervous system, according to Wolpaw, may explain why professional dancing and athletic skills take so long to develop.

For learning some simple tasks, the brain may not be necessary at all. Experiments conducted in the lab of Dr. James Grau at Texas A&M University show that the spinal cord of a rat is capable of several forms of learning even after being entirely disconnected from the brain. Among these forms are different types of conditioning, such as the one that caused Pavlov's dogs to salivate at the sound of a tone. "Conditioning" implies establishing a connection between cause and effect that one would expect to be made in the brain, yet a severed spinal cord is capable of being "conditioned" on its own. For example, a rat given a shock to its extended rear leg learns to keep the leg raised in order to avoid the shock, a form of learning known as instrumental conditioning. Amazingly, a rat whose spinal cord has been severed from the brain can still learn to avoid the shock! Such experiments suggest that simple forms of learning can be supported entirely by neural networks within the spinal cord. Furthermore, Dr. Grau and colleagues showed that this learning may depend on the same molecules as learning in the brain, molecules called NMDA receptors. When the researchers blocked the spinal cord NMDA receptors in rats, the rats were no longer able to undergo conditioning.

## Teaching an Injured Cord New Tricks

Injury presents the spinal cord with the ultimate learning challenge. The injury leaves the lower segment of the spinal cord virtually intact. The jerky, spastic movements of the body below the injury that often occur in paralyzed people can be extremely disruptive, but they indicate that the neurons in the lower segment of the cord are alive and well. However, the shrinkage or disappearance of communication channels with the brain creates a new situation, similar to that of a baby learning to walk: The damaged cord must learn to generate walking while relying on a more limited input from the brain and using different nerve circuits. In a way, after injury, the walking information is trapped inside the spinal cord, and without external help, this information may be unable to get out.

In two separate research efforts, Dr. Reggie Edgerton, working at the University of California in Los Angeles, and Dr. Serge Rossignol, at the University of Montreal, showed that the spinal cords of adult animals could be trained after injury. Both scientists had independently spent time in Professor Sten Grillner's laboratory in Sweden in the 1970s and were inspired by Grillner's studies to devote all their subsequent research to spinal cord circuitry. While Grillner had moved from the mammalian spinal cord to a simpler vertebrate model, Edgerton and Rossignol, at their respective institutions, would lead studies in adult mammals in order to make the research relevant for humans with spinal cord injury. In particular, they showed that spinally injured cats could relearn the walking patterns of a normal cat. (Such animal experiments cannot be conducted unless they obtain approval from committees composed of scientific experts and members of the public. The task of these committees is to make sure that all animal studies are well designed, humane, and aimed at achieving vital goals. In the case of spinal cord circuitry, the ultimate goal is helping humans with paralysis. "We are not testing cosmetics," says Rossignol.)

It is possible that without training, the cord of a paralyzed person loses the walking habit—a phenomenon referred to in other areas of research as "learned nonuse." Dr. Edgerton and colleagues found that spinally injured cats trained on a treadmill were three times more likely to recover stepping ability than the animals allowed to recover spontaneously. In the same study, some cats were trained to step, while others were only trained to stand. The result: The paralyzed animals were able to accomplish only the task they had been taught, not the other one. These experiments suggest that an injured spinal cord may be able to perform only the tasks it has been taught, a finding that has significant implications for humans recovering from spinal cord injury. "The system learns whatever it's experiencing; in other words, if you are in a wheelchair or remain in bed, you learn to remain in bed," Edgerton says.

When the spinal cord adjusts to postinjury reality, it seems to make the most of whatever circuitry it has left. In particular, when partially or entirely severed from the brain, it relies much more on the information it receives from the environment. Scientists are trying to determine what kind of information the injured spinal cord can interpret. Dr. Rossignol, at the University of Montreal, has found that when the cord gets no information from the brain, it becomes much more dependent on sensory cues from the legs. He has shown that in normal cats, the nerves transmitting touch information from the skin in their paws are not crucial for stepping on a treadmill, while spinally injured cats heavily rely on this touch information. In humans, this would mean that stimulation of the feet may be very important. (In fact, touch can trigger involuntary stepping in newborns: If you hold a baby and let its feet touch a flat surface, the baby will alternately lift its legs as if trying to walk. Scientists believe this reflexive response, which disappears by the age of 6 to 8 weeks, shows that the spinal machinery for walking is in place but cannot yet be controlled at will.)

Rossignol and colleagues have also discovered a chemical switch that may regulate walking. Normally, cats take weeks or even months to recover movements in their hind legs, but animals given a drug called clonidine start striding on a treadmill within minutes. The drug enhances a chemical signal that affects neurons. When the scientists blocked this chemical signal, normal cats dragged their legs and had difficulty walking. Thus, treadmill training may be one day supplemented by drugs that will speed recovery. This idea is supported by several studies with humans conducted by Rossignol's former postdoctoral fellow, Dr. Hugues Barbeau at McGill University. In these studies, clonidine improved treadmill walking in people with partial spinal cord injuries; the drug's effect was particularly pronounced in subjects with the most severe disability.

In the new treadmill therapy for humans, people with paralysis step on a moving belt while their body weight is supported by a parachute-like harness. Alternatively, the belt may remain static while the paralyzed person is propelled forward by a ceiling-mounted harness. The rationale behind treadmill training is to reproduce normal walking movements while teaching the spinal cord to interpret sensory information coming from the feet, in the hope that nerve circuitry in the injured cord will take control of walking in a new situation.

Launching the therapy in humans has not been easy. Training paralyzed people to walk is a provocative idea in itself, and it did not help that the approach is slow and labor-intensive: Every person with an incomplete injury requires the assistance of one or two specially trained therapists over several weeks or even months, while three or even four therapists may be involved in training every person with a complete injury. At the beginning, few researchers succeeded in conducting human experiments. One was Dr. Hugues Barbeau, who launched a pioneering study of 10 people with partial spinal cord injuries at McGill University. The study, completed in 1989, produced positive effects after six weeks of training, such as improved speed and quality of movement. At about the same time, Edgerton at UCLA was trying to convince his university colleagues to conduct a similar study, but nobody took him seriously. The approach, he says, involved "slaying the gods of neuroscience." Pioneering British neurophysiologists, particularly the legendary Charles Sherrington and his contemporary T. Graham Brown, had discovered the spinal cord's independent role in walking in the early twentieth century, but they attributed the stepping movements governed by the cord to simple reflexes—fixed, automatic neuronal responses. Rehabilitation professionals dismissed the idea that nerve circuits in the spinal cord could "learn."

Treadmill therapy for humans received a boost thanks to neuroscientist Anton Wernig, from the University of Bonn, who had been studying the plasticity of the nervous system in frogs and mice and became aware of Edgerton's treadmill studies with cats during a sabbatical in Los Angeles. Wernig was so impressed with Edgerton's results that upon returning to Germany he temporarily suspended his basic research and started looking for a clinic where the approach could be tried in humans. In the late 1980s he created one of the first programs for systematically training spinally injured humans to step, in a rehabilitation clinic in Karlsbad-Langensteinbach, a small town near Karlsruhe. He reasoned that although the cat and human spinal cords are different, the need for task-oriented practice must be a general feature of the nervous system in all species. "If you want to relearn walking, you need to intensively practice walking, not something else," Wernig says. "You don't practice playing the violin if you want to become a piano player." In 1995, he reported on a study in which some 150 people with incomplete spinal injuries were divided into two groups: one group received conventional rehabilitation therapy, while the other went through the innovative training on a treadmill. (Wernig refers to the moving-belt exercise device by its German name *Laufband* to avoid using the word *treadmill*, which to him is associated with the medieval machine that was operated by forced labor.) In this study, 87 percent of the people in the Laufband group learned to walk independently, while in the conventional therapy group this number stood at 34 percent.

Today Professor Wernig runs programs in Bonn and in Karlsbad-Langensteinbach, which by 2001 had put more than 600 people with incomplete spinal cord injuries through treadmill training, including some people who had been confined to a wheelchair for many years. One of Wernig's success stories, described in *Science* in 1998, was 27-year-old Thorsten Sauer, who had been in a wheelchair for six years after partially wrecking his spinal cord in a motorcycle accident. Upon completing Wernig's 10-week program in 1996, he was able to walk slowly on a treadmill without a harness while grasping parallel bars. "It was amazing," Sauer told *Science*. He learned to move about using a wheeled walker and perform such activities as negotiating a few steps and entering a narrow doorway.

Bolstered by encouraging reports from Professor Wernig's clinic, in the mid-1990s Edgerton was finally able to launch a treadmill program for humans at UCLA. While Wernig accepts only individuals with incomplete injuries, Edgerton and colleagues also study people whose spinal cord injuries are complete. These people may be unable to learn how to walk independently, but they may benefit from the training because the therapy gets their bodies in better shape. Besides, they provide researchers with a unique opportunity to study the spinal cord's independent performance.

Meanwhile, new insights into the walking circuitry in humans have been accumulating. In 1994, neurophysiologist Blair Calancie, at the Miami Project to Cure Paralysis, reported on the case of a man who had been paralyzed for 17 years by a partial injury to his spinal cord. The man, who started intense workouts in the Miami Project's rehabilitation facility, complained that he could not sleep at night because his legs started "walking." The puzzled scientists found that the man could not initiate the movements at will—even though his spinal cord was not completely cut, the remaining fibers were not sufficient to order his legs to move. The "walking" began when the man was lying on his back, and the movements appeared to be triggered by the placing of the hip joint at a particular angle. Apparently, sensory input from the hip—or possibly the pain impulses caused by inflammation that had developed inside the joint, activated the walking machinery in the spinal cord.

Another study, reported in 1997, suggests that the walking circuitry in the human spinal cord is capable of responding to the environment without "consulting" the brain. Edgerton and his colleagues at UCLA placed electrodes on the lower body and legs of able-bodied and of paraplegic subjects and changed the load on their legs while the subjects walked on a treadmill. The electrodes recorded the spinal cord's output, as measured by electrical currents associated with muscle activity. The researchers found that the pattern of changes in the output was the same in able-bodied people as in those with completely cut spinal cords. This suggested that the spinal cord, all by itself, was able to interpret load information during walking.

If a central pattern generator does exist in humans, where in the spinal cord is it located? Professor Volker Dietz, at the University Hospital Balgrist in Zurich, who studies the effects of treadmill training on spinally injured humans, believes the walking circuitry is distributed throughout the cord. This may explain why individuals with higher injuries sometimes have an easier time regaining walking than those with injuries to the low back: In higher injuries, the entire pattern generator in the spinal cord may be intact. The fact that swinging of the arms assists walking also supports the idea of widely spread walking circuits.

One way to probe spinal cord circuitry in humans has been to stimulate the cord with electrodes placed under the skin. Neurophysiologist Milan Dimitrijevic, who worked for several years at the Baylor College of Medicine in Houston, Texas, has reported with colleagues that such stimulation in the area of the low back produces stepping movements in people with complete, long-standing spinal cord injury. Not all scientists agree that these findings demonstrate the existence of a pattern generator in humans, but electrical stimulation may prove useful for training spinal cord circuitry. Dr. Elena Shapkova, of the Phthisiopulmonology Institute in St. Petersburg, Russia, has reported at scientific conferences that her team has treated some 30 paralyzed children and youngsters. According to these reports, after two to three months, a daily regimen of electrically stimulated stepping followed by treadmill therapy led to significant improvement in mobility in 24 young people. The effectiveness of electrical stimulation, however, needs to be confirmed in larger trials.

Several research centers in the United States, Canada, and Europe now have treadmill programs for research and rehabilitation— some of which have started trying the approach in people recovering from stroke, not only spinal cord injury—but the number of such programs is still small because the therapy requires special equipment and qualified personnel. Putting a person on a treadmill does not necessarily expose the spinal cord to a learning experience, and the approach can be ineffective if provided by therapists who are not versed in the new findings on spinal cord circuitry.

"Only a few clinics that advertise Laufband therapy in fact have properly trained therapists or the correct equipment," warns Professor Wernig. To make the treadmill training available on a mass scale, UCLA, together with the National Aeronautics and Space Administration's Jet Propulsion Laboratory in Pasadena, California, and the research team of Professor Dietz in Zurich, set out to develop robotic stepper devices that may replace up to four therapists. These devices will consist of a treadmill with robotic arms and a harness that will support the person and guide his or her legs on the treadmill. NASA became involved in the project because the stepper may someday help astronauts train their bodies during prolonged space missions, such as extended stays on the space station.

Personnel and equipment requirements, however, have not been the only factors precluding the use of treadmill therapy in many medical centers. The idea of "reeducating" the spinal cord is still seen as controversial: Many experts are not convinced treadmill walking is better than more conventional approaches, or for that matter, that it indeed alters nerve circuitry in the spinal cord. People with paralysis often get little exercise, so that any physical activity might conceivably improve their mobility by simply training their muscles. Two large-scale trials sponsored by the U.S. National Institutes of Health were launched in 1999 to optimize training regimens and resolve the controversy over the effectiveness of treadmill training. One trial, conducted at six medical centers in the United States and Canada and coordinated by UCLA, is scheduled to involve 200 spinally injured people over a period of five years. The other trial, designed to include 300 paralyzed people during five years, has begun at the Miami Project. Both trials include a control group receiving traditional physical therapy for comparison with treadmill training.

## Behind the Scenes of Recovery

What happens to the spinal cord circuitry during recovery is still somewhat mysterious. Most people recover some functions in the first few months after spinal cord injury and sometimes much later than that, but for obvious reasons, it is impossible to remove the cord of a living person for an examination under a microscope. And in any event, human injuries occur unexpectedly so that researchers never have the kind of "before" and "after" pictures they use in animal experiments to study postinjury changes. Among nonscientists, "miraculous" recoveries from spinal cord injuries are often credited to willpower; strong-minded people, it is said, can "will" their paralyzed legs to walk. In reality, willpower may play a role in the person's determination to seek out (or even devise their own) new rehabilitation approaches that may not be routinely recommended by medical professionals, but it is unknown whether the state of mind has an effect on nerve fiber growth. Recovery probably has more to do with the spinal cord's intelligence and plasticity, its ability to adjust to a new, postinjury situation.

Scientists have proposed several mechanisms as probable contributors to recovery of function, all of them slow, which may explain why spinally injured people take many months to improve. One likely mechanism is the sprouting of tiny offshoots from intact nerve fibers near the injury site, the same sprouting that remained controversial well into the 1980s. The sprouting of fibers, which resemble new twigs on a pruned tree, is now a well-accepted phenomenon in animals, and it probably also occurs in humans. Sprouting may sometimes be harmful: It is probably in part responsible for spasms and chronic pain after spinal cord injury. Moreover, reconnecting the two severed ends of the spinal cord by tiny sprouting fibers is unlikely to restore full function: For this purpose at least some fibers would have to undergo "true" regeneration and grow over long distances. However, sprouting is today believed to be an important compensating mechanism, and scientists think that certain types of sprouting may help people get back on their feet. Since the nerve circuits may not have to be restored exactly to their preinjury shape, sprouting may create new nerve circuits that will make up for the lost ones.

Reorganization of nerve circuitry is, in fact, another mechanism probably underlying recovery of function after spinal cord injury. Thanks to such reorganization, a small number of axons may do the job previously performed by the full complement of fibers in the intact spinal cord. Some axons may take over from others, and new nerve connections may be formed. This process probably occurs spontaneously to some degree, and treadmill training is an attempt to make it more efficient.

Yet another mechanism that may contribute to recovery is recoating spared nerve fibers with myelin. Myelin-making cells may step up their activity after injury to restore the myelin sheath of some denuded fibers, enabling these fibers to resume the conduction of impulses. Finally, a controversial and still unproven concept is that of spontaneous regeneration. Several scientists have expressed the view that, against all odds, some regeneration may take place in the human spinal cord. If it indeed occurs, such regeneration could account for the fact that paralysis typically retreats by an inch or two down the spinal cord in the first few months after the injury, allowing some people to regain a certain degree of function.

Treadmill training may barely scratch the surface in teaching the spinal cord to make the most of its own compensatory mechanisms. In the future, such teaching may become a vital part of regeneration approaches. "Otherwise, you may have 100 percent regeneration and not be able to walk," says UCLA's Reggie Edgerton. Training may not restore normal walking, but it may help people recover a degree of mobility that can make a huge difference in their lives. Moreover, it may get some people out of wheelchairs even when their brains and spinal cords communicate via a spindly thread of a few regenerated fibers. "This makes the regeneration story much more realistic," Edgerton says.

# 16

# The 2-Million Quest

If spinal cord regeneration proves effective in humans, will it benefit only future victims of spinal cord injury? What about people who have been paralyzed for years? "Curing Christopher Reeve" has become a catchphrase for curing paralysis, but what are the chances that Reeve, and the other 2 million people worldwide now living with spinal cord injury, will benefit from advances in regeneration research?

Only a few years ago, the vast majority of scientists would not direct any efforts toward long-standing, or "chronic," spinal cord injuries. Fresh injuries—in medicine, the formal term is *acute injuries*, which sounds like "extremely serious" but actually means "in the process of occurring" or "still an emergency"—are challenging enough and still poorly understood. Moreover, it makes sense to try new therapies first in experimental animals with acute injuries because such studies are faster and cheaper. But now the number of laboratories designing animal studies to investigate chronic injuries is increasing slowly but steadily. The increase is a sign of growing confidence. Some leading laboratories in the field are getting such good results with regeneration in an acute setting they feel confident about tackling the more knotty chronic situation. (In humans an injury is usually considered chronic a year or two after the initial trauma, by which time all the trauma-related processes in the cord subside, as does the process of spontaneous recovery.)

It takes a particular temperament to go into chronic injury research. While regeneration studies are always painstakingly slow, waiting for the injury to become chronic before beginning the experiment means an even greater delay in obtaining results. "These experiments are not suitable for people who want results faster; you just can't rush them," says John Houle of the University of Arkansas for Medical Sciences in Little Rock, a pioneer of chronic injury research in animals. Temperament requirements aside, chronic injuries are difficult to work with for scientific reasons. Shortly after the injury, many neurons die and many axons of surviving neurons retract; then the cascade of secondary injury destroys additional nervous tissue. These events greatly increase the damage that needs to be repaired. A cyst, a fluid-filled cavity, often forms at the injury site. Moreover, with time, the site of injury fills with scar tissue, which creates a chemical and physical barrier to regeneration. Also, immediately after the injury, the cut or crushed axons attempt to repair themselves, but as these attempts subside, extra effort is needed to reawaken regeneration. Then there is the question of money: Paying trained technicians to care for paralyzed animals for extra weeks or months sharply increases the cost of experiments. Scientists are under constant pressure to produce fast results at the lowest cost, and many cannot afford labor-intensive chronic injury studies. It is for these reasons that, for the most part, scientists try to repair the spinal cord of a rat or mouse immediately after damaging it. This focus on acute injuries makes it possible to screen a greater number of potential approaches more efficiently. Chronic spinal cord injury experiments are usually done only after the acute studies show promise. Most laboratories conduct chronic experiments in parallel with acute, and Houle's small lab is the only one in North America focusing exclusively on chronic injuries.

Dr. Houle, a low-key, soft-spoken native of Cleveland, started studying chronic spinal cord injuries in the mid-1980s, when the possibility of regeneration even in acute injuries had barely been established. He first investigated the topic while working as a postdoctoral fellow in the laboratory of Paul Reier at the University of Florida, then continued after setting up his own lab in Little Rock in 1987. It is not by chance that he chose to do his research away from the limelight and from the high-powered environments of larger American cities and schools, despite the difficulty in attracting students to a small university in the rural state of Arkansas. The unhurried pace of the American South seems to suit a line of research that demands persistence and patience: "I like the pace of life here, and I like working in a small institution where people are not too spread out and have time for one another. No one is pushing me, I am able to sit and think things through without being forced to change all the time." Even though more laboratories are now venturing into research on chronic injuries, Houle is not afraid of competition. Nobody, he says, has as much experience in working with these injuries as he does.

In 1991, Houle was the first to demonstrate that spinal cord axons can regenerate long after an injury. He created an injury in a rat spinal cord, waited a month, and then performed peripheral nerve transplants, the same as those done in Albert Aguayo's laboratory in Montreal a decade earlier. Despite the delay, some of the cut axons in most major tracts of the cord retained the ability to regrow. In subsequent experiments Houle extended the time lapse between injury and treatment to six months—and, surprisingly, some of the rats' neurons still regrew their axons.

One of the main topics in Houle's laboratory is the study of changes that occur in nerve growth after injury. He found that in rats, the regeneration capacity of the cut axons steadily diminishes with time, but it can be boosted or revived with growth factors. Unfortunately, no single injection of one growth factor would do the trick. Not only do different neurons need different growth factors, but Houle, along with other investigators, found that the situation is even more complicated. The responsiveness of the same neuron changes during the postinjury period, so that a growth factor that caused the cell to regrow its axon early on may no longer be effective at later stages. A fuller understanding of the cells' evolving needs may help stimulate regeneration long after the injury has occurred.

## Exaggerated Death Reports

In some cases, chronic injury studies yield pleasant surprises. Reports about the death of certain neurons after injury had been greatly exaggerated, scientists at the University of British Columbia discovered. Injured neurons of a particular type had been considered dead because they undergo massive shrinkage after their axons are cut, but in fact they had simply shrunk so much they could no longer be detected. When the scientists, led by Wolfram Tetzlaff, added a growth factor to these neurons in the brains of spinally injured rats, the invisible cells "came back" and their normal numbers were restored. Moreover, a full year after injury the resuscitated cells regrew their axons, the team reported at the 2000 annual meeting of the Society for Neuroscience. "This is good news because it shows that in a chronic injury, where we believed in the past the cells were dead, they may still be there and we may have the tools to revive them," Tetzlaff says. Previous studies indicate that having its axon cut off doesn't always kill the cell; sometimes, on the contrary, the cut may spur the cell to produce substances needed for regeneration. It seems that a great deal depends on the site of the cut. Scientists have discovered that if the axon is severed close to the cell body, leaving only a short stump, the cell is likely to die or shrink; conversely, if the cut is too far away, the cell body may not be galvanized into action. Dr. Tetzlaff is following up on these findings, trying to identify the genes that increase the capacity of injured neurons, particularly the chronically injured ones, to regenerate.

Another surprise, dubbed by the scientists "a perverse finding," came from Georgetown University. Working with rats, a team headed by Barbara Bregman wanted to define the time limit of a treatment's effectiveness after injury. The researchers started with a four-to-eight-week interval after the injury, convinced that after such a long wait the approach, consisting of growth factors and transplants of fetal spinal cord tissue, would not work, and they intended to decrease the interval gradually. Much to their amazement, when the treatment was given four to eight weeks later, it was more effective than in acute injuries! Dr. Bregman, whose laboratory expanded its studies in the late 1990s to include chronic injuries, proposes several explanations. One of them is a "conditioning" effect known from peripheral nerves: The initial injury might have aroused the neurons, making them more responsive to the treatment given several weeks later. One of Bregman's research goals is to reveal the differences between the responsiveness of acutely and chronically injured neurons.

Other laboratories are studying complications occurring in chronic injuries or testing in a chronic setting new therapies that were initially developed in acute injuries. Scientists at the University of California at San Diego, after successfully regenerating freshly injured axons using growth-factor gene therapy, tried the same approach three months after the injury. In 1997, the team, led by Dr. Mark Tuszynski, reported that in rats with the chronic injury the axons had retained their ability to regrow. Other landmark approaches developed with acute injury—Martin Schwab's IN-1 antibody, the Cheng–Olson bridge therapy, Michal Schwartz's macrophage therapy—are also being tested in animals with chronic injuries.

What do all these studies mean for humans with chronic injuries? Scientists can regenerate the nerve fibers of a rat six months or even a year after injury, but how does this translate into months or years of human life? A cautious approach is to compare metabolic and heart rates. These are about four times slower in humans than in rats, so six months in a rat could be equal to some two years in a human. A more optimistic approach is to use percentages of the life span: Six months account for about one-sixth of a rat's life, so this period could theoretically be equivalent to about 15 years in humans.

While such extrapolations are speculative, some findings in humans are encouraging. Decompression surgery, which frees the spinal cord from adhesions to its membranes or from other mechanical obstructions, can produce improvement in function years after the injury. Paralyzed people occasionally report that they spontaneously recover small movement in their muscles long after the injury. Such improvements are typically minor and make little difference to the person's total function—he or she may, for example, regain the ability to move the big toe—but these reports support the notion of long-term recovery potential. There are also promising findings at the molecular level. At the University of California at San Diego, scientists found that 15 years after injury, the genes for growth factor receptors continue to be active in human spinal cord fibers, suggesting that these fibers may respond to growth stimulation.

All the usual caveats of regeneration research hold for chronic injury experiments: The number of studies is still small and no one knows if the therapies that work for rats will apply to humans. However, studies suggest that, even years after injury, all hope need not be lost. Some of the impending human trials will enroll people with chronic injuries and may provide the long-awaited answers.

# 17

# Man–Machine Mergers

Years before spinal cord regeneration for humans was on the horizon, researchers started working on neural prostheses that would allow people with paralysis to perform simple tasks that able-bodied people take for granted: clenching a fist, emptying the bladder, walking. Like the mother of all prostheses, the wooden leg, neural prostheses are designed to replace a missing or malfunctioning body part, in this case, a network of nerves. By the early twenty-first century, several of these devices became commercially available, while numerous others were in various stages of research and development.

Will neural prostheses solve the problem of paralysis, making the regeneration quest obsolete? Probably not. It is unknown whether the prostheses will ever be able to produce versatile natural movements, fulfilling the mythical merger of man and machine prophesied by science fiction. Besides, given the choice, paralyzed people would prefer having their own body back, not a mechanical substitute. However, technology may supplement biology, offer interim solutions, or provide relief in the most stubborn cases. The field of neural prostheses already has one major success story: Cochlear implants, approved by the U.S. Food and Drug Administration in 1986 and placed in the ears of more than 30,000 people worldwide, successfully restore hearing in individuals with certain types of deafness by stimulating the auditory nerve. Systems that may partially compensate for loss of vision are also being developed. In paralysis, neural prostheses may supplement regeneration: For instance, if the axons regenerate but fail to form appropriate connections, an implanted array of microelectrodes could help direct them to the right targets or compensate for connections that turn out to be wrong. It may also be possible to use an implanted prosthesis to amplify the signals transmitted from the brain to the muscles by regenerated fibers, in case the number of these fibers is small and the signals are weak. Other types of prostheses may restore a degree of normality to the lives of some people with severe disabilities before biological studies bear fruit.

The most extreme form of paralysis is the so-called locked-in syndrome: Sufferers are fully conscious but so severely paralyzed by injury, stroke, or neurological disease that they cannot move a single muscle. One famous case has been that of Jean-Dominique Bauby, former editor in chief of the French *Elle* magazine, who was able only to blink his left eye after being locked in by a brain-stem stroke at the age of 43. By blinking at appropriate letters when the alphabet was displayed to him, Bauby "dictated" a critically acclaimed book, *The Diving Bell and the Butterfly*, which was published in France in 1997, two days before he died of a heart attack. Some individuals with locked-in syndrome, however, cannot even blink, and for them neural prostheses may offer the only means of communicating with the outside world. Such prostheses, for lack of other options, must be operated by a practical version of wishful thinking—thought focused on a particular operation. This may sound futuristic, but as early as the 1960s scientists discovered that people can be trained to control certain electrical signals emitted by their brains. Now researchers are applying this finding to developing neural prostheses whose purpose was aptly summed up in a *Science* headline: "Turning Thoughts Into Actions."

The simplest operation that can be driven by thought, moving a cursor on a computer screen, is already being performed by a handful of human subjects in several studies. In two research projects in the United States and Germany, locked-in people are trained to "unlock" themselves by controlling their electroencephalograms, or EEGs, the pattern of brain waves recorded by electrodes placed on their scalp. While the individual thinks of a cursor moving in a particular direction, the electrodes pick up the thought and cause the cursor to move on the screen. In the course of numerous training sessions, people can learn to answer yes–no questions or slowly spell out words. A more effective but also more invasive way of picking up neural signals is to implant electrodes directly into the brain. Scientists at Emory University in Atlanta, Georgia, were the first to perform this feat in locked-in humans. Their first research subject died of her disease shortly after surgery, but the second one, who received the implant in March 1998, learned to spell messages on a computer. The 53-year-old man, unable to move since his 1997 brain-stem stroke, was trained to compose phrases and respond to questions by "willing" the computer cursor to move on the screen.

Multiple electrodes implanted into the brain can record enough neural signals to mind-control even more complex operations, such as moving an arm. This line of research began with basic studies of movement control by the brain, but when technological advances made it possible to record from dozens of neurons at once and scientists began to decipher the neuronal patterns that make movement happen, the idea was born to apply the findings to neuroprosthetics.

One fascinating prototype system for a thought-controlled prosthetic arm was tried in 1999 in monkeys at Arizona State University in Tempe. Scientists and engineers working in collaboration with the Neurosciences Institute in La Jolla, California, have been using an array of implanted microelectrodes to record the firing of neurons that occurs when a monkey intends to move an arm. A computer translates this firing pattern into instructions for a robotic arm, which mimics the animal's movement. At the beginning of these experiments, the robot was in a separate room and the monkey was unaware it was being "aped" by a machine, but then the scientists developed a program for training the monkey to consciously control the robotic arm by thought alone, without making movements in parallel. In the same manner, a person with severe paralysis could be trained to control a prosthetic arm purely by intention.

The science behind such prostheses is already established, and making them available to humans is only a question of better engineering, says Professor Andrew Schwartz, head of the Arizona team. One technological challenge is the need to record from a greater number of neurons: Schwartz's team taps into the neural chatter of 30 to 50 cells, but recording from 100 to 200 neurons would be required for effective arm control. The other issue is reliability. Large electrodes, such as those used for muscle stimulation, can work reliably for years, but recording electrodes, which must pick up tiny signals from nerve cells, often do not last. "If we could record from 100 neurons reliably, we could implant this in humans right away," Schwartz says. He expects clinical trials to begin around 2005 and envisages a neural prosthesis in the form of a humanlike arm that could be attached, say, to a wheelchair, and "willed," via a brain-implanted chip, to perform daily chores.

## A Bridge to the Future

Other, more "down-to-earth" neural prostheses are already benefiting people with paralysis. Jim Jatich, 51, who had been unable to use his hands since the 1977 diving accident that left him paralyzed, helped engineers develop a hand prosthesis called Freehand and became its first recipient. "I'm using my hand again. I'm picking up a fork to feed myself, and picking up a pen to write again," Jatich told *Technology Review* magazine in the spring of 2000. "That's a big emotional change in my life." Freehand allows people with tetraplegia, who can use their arms and shoulders but are unable to open or close their hands, to grasp objects so as to perform

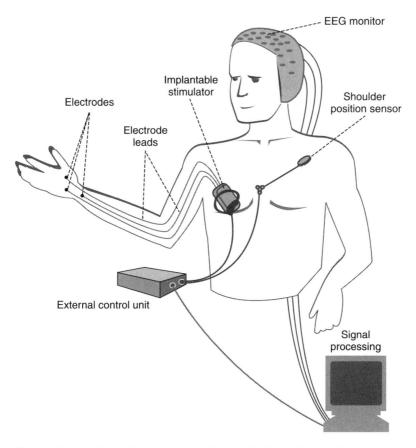

Control of an implanted hand prosthesis, Freehand. (NeuroControl Corporation.)

such tasks as eating, grooming themselves, or inserting a floppy disk into a computer. The system falls into the FES, or functional electrical stimulation, category of prostheses, which stimulate muscles to contract by applying electric currents. FES is widely used to exercise muscles during rehabilitation, but in the exercise setting the electrodes are placed over the skin. In contrast, in many FES prostheses, electrodes are implanted under the skin and activated by radio signals to orchestrate muscle movement. Freehand, recently approved by the U.S. Food and Drug Administration and used in about 200 people around the world, consists of long wires implanted in the arm and terminating with platinum electrodes placed next to the nerves feeding the hand muscles. The system, developed at Case Western Reserve University and the Cleveland Department of Veterans Affairs Medical Center, is activated via an external control unit by movements of the opposite shoulder: By shrugging, say, the left shoulder in a particular way, the person can open and close the right hand.

The neural prosthesis implanted in the largest number of people with paralysis is the so-called bladder pacemaker. Just as the heart pacemaker causes the heart muscle to pump blood, the bladder pacemaker can contract the bladder on demand, prompting it to expel urine. Some 2,000 people in 20 countries have been implanted with this type of prosthesis, commercially available in the United Kingdom since 1982 and approved by the U.S. Food and Drug Administration in 1998. Other FES devices that are available or under study include systems that assist breathing and coughing, improve circulation, control spasticity, produce erection and ejaculation, and treat pressure sores. Several centers worldwide are developing FES prostheses for standing and transfers. These systems, already implanted in several dozen volunteers, stimulate the leg muscles to stiffen the trunk, hips, and knees in order to allow a paralyzed person to stand for limited periods of time, say, in the courtroom or classroom, to reach objects on high shelves, to maneuver into spaces where wheelchairs won't fit, or to be transferred easily between a wheelchair and a bed or toilet. Prostheses for walking are also under development, but they are likely to take longer to reach the market than the standing and transfer systems. Muscle fatigue is a major problem with walking, as the prostheses rely on prolonged contraction of muscles to maintain stability and move the body forward. Other major issues that need to be resolved before walking prostheses can be widely used are versatility, for example, adjustment to walking on different surfaces, and overall reliability of the external and implanted components. "One of the critical aspects of any prosthesis is that you want to forget about it," says Dr. William Heetderks, head of the Neural Prosthesis Program at the U.S. National Institutes of Health. "You want to know it's going to work today, tomorrow, and next year."

The potential advent of regeneration therapies dictates special care in installing the prostheses, particularly the ones that require placing several feet of wires under the skin. "The approach we've taken to installing our system leaves all the anatomy intact," says Dr. Ronald Triolo, head of a team working on a standing-and-transfer prosthesis at the Cleveland FES Center, a consortium of medical centers with a broad research program in neuroprosthetics. "The tendons are not cut or rerouted, so that if regeneration therapies become available to heal the damaged spinal cord, the anatomy will still be there and the person will be healthier because they've been exercising."

The need to implant long wires under the skin may be eliminated by a new generation of neural prostheses—microchips to be implanted directly into the spinal cord. The approach, known as microstimulation because it relies on tiny, hair-thin electrodes passing minute electric currents, is in early stages of development in laboratory animals by several research teams. If the studies show it to be safe and effective, microstimulation would involve placing a microchip directly into the spinal cord and activating this chip by radio signals that would replace brain-derived commands for different tasks: emptying the bladder, standing, or walking. "The spinal cord contains all the circuitry for specific functions, and the idea is to access this circuitry instead of activating individual muscles," says Dr. Vivian Mushahwar of the University of Alberta in Edmonton, Canada. Mushahwar and her colleagues on the Edmonton team headed by Dr. Arthur Prochazka managed to trigger coordinated stepping movements in healthy cats by stimulating six to twelve microwires implanted in the spinal cord. The cats appeared to experience no discomfort and the implanted electrodes remained in place for six months, even though the animals were allowed to run around.

In the future, it may also be possible to combine different types of prostheses. Thus, if brain implants can be combined with FES prostheses, the technology may allow people with little or no body movement to mind-control their limbs. At the Cleveland FES Center, Freehand recipient Jim Jatich has already learned to activate his prosthesis with the help of electrodes attached to his scalp, which translated his thought into movement. "I got tears in my eyes, turned to my sister, and said, 'Damn, I actually moved my hand by thinking about it,' " he told the *Technology Review* reporter. External electrodes allowed Jatich to perform only very crude movements, but an implanted chip could do better, making it possible to effectively control implanted hand or leg prostheses.

Combined devices of this sort, however, if and when they become a reality, may serve as stopgap measures. Replacing brain circuitry with electrodes gives scientists a unique opportunity to study the human nervous system, but it is unlikely to become the preferred medical procedure. "In the long run, I doubt that implanting a bunch of hardware in your brain will be the optimal solution for spinal cord injury," says Arizona State University's Professor Schwartz. "Years from now, I think this body of work will be valued for what we've learned about the science, not necessarily for making a better arm for a person with paralysis. By then biological approaches to spinal cord repair will be better than what we can do. But neuroprosthetics can be useful for a while, as a bridge to the time when biological approaches become viable."

# Epilogue

For about five months after her June 2000 accident, Melissa Holley, who had received immune regeneration therapy in Israel, could move no muscle in her lower body. Then, on Thanksgiving night, while sitting in her bed at the Craig Hospital in Denver, Colorado, where she was undergoing rehabilitation after returning from Israel, she felt a spasm in her left leg and suddenly discovered that she was able to move her thigh a fraction of an inch. She pressed a call button and asked a nurse to send in her mother, who was in a nearby room watching a movie. "My nerves were shooting off," Melissa would later tell a local reporter. "My left leg suddenly bent and came up toward me. I started bouncing it in a rhythm. I didn't know what to do, I was so happy. I had tears running down my cheeks. [My mother] was like, 'What's wrong?' I was like, 'No, Mom, this is good.' I put her hand on my thigh and she could feel the muscle contraction."

When this book went to press in 2001, Melissa had recovered

most of the sensation in her lower body and could move her toes, lift her feet, and flex some of her thigh muscles. She was planning to start learning how to get around on crutches while wearing braces on her legs. However, it was too early to know the full extent of her recovery, and major questions remained to be answered. Had Melissa's spinal cord fibers regenerated thanks to the treatment? Or would she have recovered anyway? Medical miracles are rare, but they cannot be excluded. Moreover, a small proportion of people originally diagnosed as having a "complete" spinal cord injury eventually recover a certain degree of feeling below the injury site; in rare cases, they recover some, usually unsubstantial, muscle movement. In an encouraging development, in the spring of 2001 physicians running the Israeli trial reported that two other people who had received the immune therapy showed signs of recovery. Both had recovered sensation in the formerly paralyzed parts of their bodies, and one, paralyzed from the neck area down, had also recovered movement in his fingers. But even if these hopeful signs foreshadowed success, it would be impossible to know, on the basis of these cases alone, whether the immune therapy was effective. More people would have to be treated before doctors could draw definitive conclusions. As for other regeneration therapies, at the time of this writing most were still not ready to be tested in humans, and the few sanctioned regeneration trials that had already begun had not yet been completed.

In writing about spinal cord regeneration and repair before the search has culminated in a happy ending, I yearn to know how this story will unfold. In a way, writing about the field now is much as it must have been writing about DNA and heredity in the years preceding the discovery of the double helix. Only after a great scientific discovery is made does the newly revealed truth appear amazingly obvious; today it seems as if the role and structure of DNA should have always been known to researchers, rather than uncovered only some 50 years ago. And only in retrospect do the milestones preceding a discovery arrange themselves neatly in a line, like stepping stones leading to a pedestal.

It is possible that one day this is how we'll be talking about recovery from spinal cord injury. If different therapies are proven to be successful, it will be possible to chart the major contributions that led to each one and to show how the different lines of research ran in parallel or converged. But this is something that people analyzing the field will be able to do in the future. For the time being, the suspense is still there. Will any regeneration therapies work for humans? If so, which ones? Will scientists one day indeed change the verdict of the famous 3,500-year-old Egyptian papyrus that labeled spinal cord injury "a disease that cannot be treated"? The feat still sounds like a miracle, but now there is real hope that eventually they will; and then medical textbooks—and this book too—will have to be rewritten.

# Appendix:
# The Spinal Cord Before and After Injury

The distinction between the brain and the spinal cord is somewhat artificial as the two organs are made of the same cells and together form one system, called the central nervous system. Bony armor protects both: The brain is shielded by the skull, and the spinal cord—an extension of the brain stem, the lowest and most ancient part of the brain—is encased by the vertebral column. The cord is enveloped in three layers of connective tissue coverings called the **meninges**. The toughest, outermost layer of the three is called the **dura mater** (from the Latin words *dura* for "tough" and *mater* for "mother").

The spinal cord's shape in different species is affected by the size and importance of the limbs. The human cord, 16 to 18 inches long, is about three-quarters of an inch in diameter but it is thicker in two places. The upper thickening, called the **cervical enlargement**, contains nerve cells that communicate with the arms; the thickening at the level of the lower back, called the **lumbar enlargement**, communicates with the legs. The whale, in contrast, with its massive trunk and only rudimentary limbs, has a large cylindrical cord without enlargements. The dinosaur, with its massive lower limbs, had an enlargement in the lower part of its spinal cord that rivaled its brain in size.

Two major systems of nerve cells operate in the spinal cord. **Descending,** or **motor pathways** send commands from the brain to the body to control the muscles and to supervise the autonomic nervous system, which in turn communicates with the heart, intestines, and other organs. **Ascending** or **sensory pathways** transmit signals from the skin, muscles, and internal organs to the spinal cord and then up to the brain.

The cord, like the brain, is made up of nerve cells, three types of support cells known together as the **glia** (**oligodendrocytes, astrocytes,** and **microglia**), and blood vessels. The inner part of the cord, butterfly-shaped on a cross section, is the **gray matter**, which integrates and processes information. The outer rim of the cord is the **white matter**, which owes its color to the large proportion of nerve fibers covered with the fatty protective sheath called **myelin**.

Most nerve cells in the gray matter are **interneurons**; they are responsible for communications inside the cord. Some of the larger nerve cells are **motor neurons**; they project their nerve fibers, or axons, outside the cord and send movement commands directly to the muscles. In humans, the cell bodies of motor neurons that activate the muscles of the arms and legs reside in the cervical and lumbar enlargements, but there are motor neurons in other parts of the cord as well; the ones in the C3 and C4 segments, for instance, control muscles involved in breathing. The cell bodies of **sensory neurons**, which carry information from the body into the spinal cord, are located immediately outside the cord in clusters of nerve cells called **dorsal root ganglia, or DRGs**.

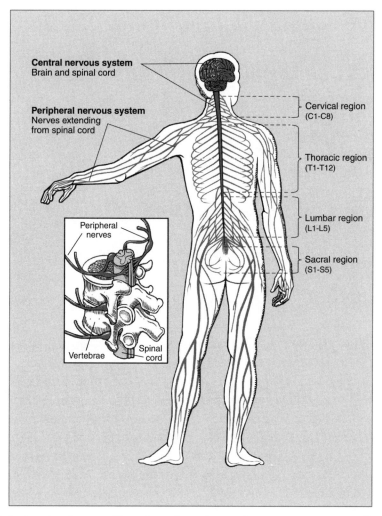

Spinal cord and nerves. (From *Brain Facts*, Society for Neuroscience, 1997. Illustration by Lydia Kibiuk.)

The white matter on the outside of the cord contains nerve fibers of different length. Some of the fibers connect the spinal cord with the brain, while others serve as links between different segments of the cord. Thus, although the white matter looks like a uniform milky layer, it is in fact a complex mesh of fibers that have different origins and different destinations. The tips of axons dip

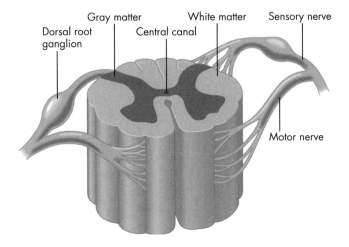

Diagram of a cross section through the spinal cord. (From *Biological Psychology*, 7th edition, by J.W. Kalat, 2001. With permission of Wadsworth Group, a division of Thomson Learning.)

into the gray matter, where they make connections with interneurons or motor neurons that relay their signals. Some signals go through several relay stations before they reach their destinations. For example, movement commands from the brain may be relayed by numerous interneurons before they reach motor neurons, which convey the signal to the muscles via the spinal nerves.

Exiting from the cord are the "wires" that connect it with the body, 31 pairs of **spinal nerves**. These nerves are "mixed," in that they are attached to the spinal cord by two bundles of nerve fibers: The bundle attached on the side of the back carries sensory information to the cord; the bundle closer to the stomach sends out motor commands to the muscles as well as signals to the autonomic nervous system. The two bundles come together a short distance after exiting the cord, but then the spinal nerves immediately divide into several branches that extend upward and downward. Although the spinal cord is not visibly segmented, each portion of the cord that gives rise to a pair of spinal nerves is referred to as a spinal **segment**. At its lower end, the cord ends in a bunch of nerves that bear an uncanny resemblance to a horse's tail, hence their Latin name, **cauda equina**.

## Types and Consequences of Injury

Two major factors determine the consequences of spinal cord injury: its extent and the level at which it occurs. In the past, an injury was considered **complete** if the spinal cord was actually cut in two, but in the vast majority of cases, the cord is bruised or crushed rather than cut. Now injuries that entirely interrupt communication with the brain are called complete; people with such injuries have no voluntary movement or sensation in their bodies below the injury. When the spinal cord is partially damaged, the injury is referred to as **incomplete** and the person has limited sensation and movement in the lower body. The proportion of people with incomplete injuries, who sometimes recover substantially, has been on the increase and, in the United States, now stands at some 60 percent, probably thanks to improving postinjury care.

Doctors describe the level of the injury by the segment, or the pair of spinal nerves, that it affects. The nerves, named and numbered according to their place of origin, are grouped into several categories. Eight pairs of **cervical nerves**, C1 to C8, ensure communications with the head, neck, diaphragm, and arms. Very high injuries often lead to the loss of involuntary functions; thus, people with injuries above C4 may require a ventilator in order to breathe. C5 injuries may enable the person to control the shoulders and biceps, but not the wrists or hands; people with C6 injuries may have wrist control but no hand function. Twelve pairs of **thoracic nerves**, T1 to T12, extend to the chest and abdominal muscles. Many hand muscles are controlled by T1. People with injuries affecting T2 to T8 may have the use of their hands but poor trunk control. Five pairs of **lumbar nerves**, L1 to L5, innervate leg muscles. People with injuries at this level experience reduced control of hip muscles and legs. Five pairs of **sacral nerves**, S1 to S5, are responsible for bowel, bladder, and sexual function and control muscles in the feet, calves, and buttocks. The lowermost pair of **coccygeal nerves** exits from the coccyx, or tailbone.

## Spinal Cord Tracts

Anatomists distinguish more than a dozen different tracts in the white matter of the human spinal cord. The tracts are bundles of nerve fibers that travel together and probably share much the same functions. **Sensory tracts** keep the central nervous system informed of changes in the internal organs, joints and muscles, and in the environment, while **motor tracts** carry nerve impulses responsible for voluntary movement. The names of different tracts reflect where they begin and end. For instance, the corticospinal tract conveys impulses from the cortex down the spinal cord, while the spinothalamic tract begins in the spinal cord and carries impulses to the brain region called the thalamus.

The exact role of different tracts in humans is complex, but anatomists and physiologists do have a general idea. The most accessible to clinical examination is the **corticospinal tract,** sometimes also called the pyramidal tract, the latest to appear in evolutionary terms. All mammals have the corticospinal tract, but it is particularly developed in primates, including humans. Thanks to the corticospinal tract, primates have the most elaborate connection between the cerebral cortex, the "thinking" part of the brain, and the spinal cord. In primates, for example, unlike other mammals, many corticospinal fibers—among the longest fibers in the nervous system—make direct synapses with motor neurons, which gives the cerebral cortex a direct and powerful control over movement. The corticospinal tract confers speed and agility on voluntary movements and is believed to be responsible for fine movements of the hand and other types of skilled motion. In explaining the role of different spinal cord tracts, Maryland professor of neuroscience Jerald Bernstein, now retired, used to ask his students why an alligator cannot play the piano. The answer, among other reasons, is that the alligator has no corticospinal tract.

Fortunately, scientists may not have to reconstruct all the spinal cord tracts in order to restore function after paralysis. Thanks to the plasticity of spinal cord circuitry, after injury, spared tracts may take over from the injured ones, as described in Chapter 15.

## Spinal Cord Reflexes

Reflexes are fast, automatic, and predictable responses to stimuli that allow the body to make quick adjustments to a changing environment. Spinal cord reflexes are numerous and powerful. Best known among them is probably the **knee-jerk**, whose function during walking is to maintain upright posture by instantly correcting the tendency of the knee to bend. When the knee bends, the thigh muscle is stretched and this information is relayed to motor neurons in the spinal cord, which automatically respond by signaling this muscle to contract in order to return the knee to an extended position. During a neurological examination, this reflex is checked by a tap on the tendon just below the knee. Another example is the **withdrawal reflex**, which causes the leg or arm to withdraw instantly from the source of pain. In this reflex, sensory neurons conduct nerve impulses from pain receptors to the spinal cord; these impulses, traveling via fixed routes, are relayed through interneurons to motor neurons, which in turn stimulate muscles to contract—for example, to pull up the leg after the person has stepped on a tack.

## The Nerve Impulse and the Synapse

Although the nervous system is sometimes likened to a network of electric cables, it is radically different from electrical circuits underlying, say, household appliances. Activity within the nervous system reflects the flow of charged chemical particles across nerve fiber membranes; when neurophysiologists perform electrical studies of the nervous system, they record the electrical signals generated by the movement of these particles.

Transmission of a **nerve impulse** begins when the neuron's membrane temporarily opens up in one spot to allow the entry of sodium ions, to which the neuron is normally impermeable. Since

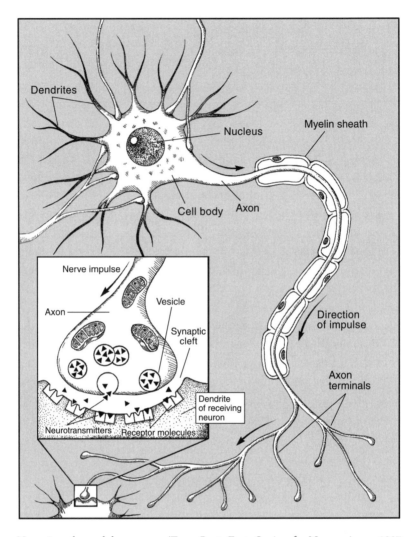

Nerve impulse and the synapse. (From *Brain Facts*, Society for Neuroscience, 1997. Illustration by Lydia Kibiuk.)

the sodium ions are positively charged, they explosively rush into the negatively charged interior of the cell. This brief inrush of positive particles is sometimes referred to as the "firing" of a neuron. The movement of the ions affects adjacent parts of the cell membrane, which in turn become permeable to sodium ions, and the process is repeated as the impulse travels along the nerve fiber. (Sodium ions do not accumulate indefinitely inside the fiber; a special mechanism continuously pumps them out.) When the outside of the fiber is covered with myelin, the nerve impulse travels faster because rather than crawling along the membrane, it "jumps" from one myelin-free spot on the membrane to another, as described in Chapter 6.

Transmitting the information carried by the impulse from one cell to another is a formidable challenge. The signal can no longer travel along the membrane but must cross a tiny space between two nerve cells, or between a neuron and a muscle cell, called a **synapse**. The synapse is arguably the most important structure in the nervous system because it enables communication—the system's main reason for existing. At the synapse, the nerve impulse resorts to a new "means of transportation": The firing neuron discharges chemical messengers called **neurotransmitters** into the synaptic space. In a way, the synapse functions like a valve: The neurotransmitters pour out from tiny bubbles in the tip of the firing axon, which release their contents when the nerve impulse invades the tip. On the other side of the synaptic space, different neurotransmitters trigger changes in the membrane of the receiving cell, potentially initiating a new nerve impulse in that cell.

In this manner, nerve cells communicate with one another and with muscle cells. This may sound like a cumbersome system of communications but in fact it is extremely efficient: A single neuron can have thousands of synapses with other cells and it can fire impulses dozens or even hundreds of times every second. Synapses, which are established between axons and the nerve cell bodies or their extensions, called **dendrites**, allow the transmission of precise information between neurons and form the basis of neuronal circuits in the nervous system. When we talk about nerve fibers being "wired" into circuits, we mean the arrangement of synaptic connections between different cells.

## What Happens in the Spinal Cord After Injury?

Except in gunshot and knife wounds, in most injuries the spinal cord is not cut across but compressed, or in technical language, "contused." The extent and course of the damage depend on the speed and duration of the compression: If the speed is less than 0.5 meters per second and the compression lasts less than 10 to 20 minutes, the cord may not suffer much permanent damage. If the compression is slow but lasts longer than 10 to 20 minutes, cells begin to die as a result of interrupted blood supply. If the speed exceeds 0.5 meters per second, the cord instantly experiences mechanical damage. Surgery is sometimes performed to free the cord from pieces of the vertebral column or other mechanical obstructions that may be pressing on nerve fibers after injury.

In addition to immediate mechanical trauma, waves of **secondary damage** continue to spread through spinal cord tissue after the injury. In the first minutes, blood leaking from broken blood vessels causes tissues to swell. Blood vessel damage and swelling interrupt the delivery of nutrients and oxygen to cells, starving them to death. Chemicals spilling from damaged cells can be toxic to adjacent cells and to nerve fibers, killing them or causing the cells to commit suicide. Prolonged inflammation and swelling further harm tissues. These processes continue for days or even weeks after the initial trauma, and while the zone of damage may be small at first, it can ultimately expand severalfold. A fluid-filled cavity called a **cyst** can form at the injury site, creating an impenetrable obstacle to any potential regrowth of nerve fibers. A **glial scar**, clusters of glial cells that multiply after the trauma, form around the cyst and in other parts of the injured tissue, releasing growth-inhibiting chemicals and creating a physical barrier to growth.

Here is how the injury affects different structures in the spinal cord:

**Nerve fibers (axons):** Axons severed by the injury snap apart like two ends of a rubber band. This happens because in the healthy cord, these axons are normally stretched: during development, bone grows much faster than do nerve fibers, so that an adult spinal cord ends up being shorter than the vertebral column. For this reason, stitching severed axons together after injury is problematic. Moreover, while the part of the axon attached to the cell body can survive, the part detached from the cell body quickly begins to disintegrate in a process known as **Wallerian degeneration**. Since about half of the nerve fibers in the spinal cord are descending and half are ascending, Wallerian degeneration kills fibers both below and above the injury. The fibers that did not cross the injury site, the ones that run their full course on either side of the injury, may remain intact and form live circuits. However, when communication with the brain is interrupted, circuits above the injury site no longer receive sensory input from the lower body, while circuits below the injury site no longer receive movement commands from the brain. Nerve cells in these circuits probably form new connections to replace the lost ones. Such new circuits forming above the injury are believed to be responsible for the pain that some paralyzed people experience, while random new connections below the injury—coupled with the loss of suppressive mechanisms—may account for spasms.

**Myelin:** Many axons are not cut or crushed by the injury, but they no longer conduct impulses properly because they partly lose myelin, their insulating fatty sheath. Oligodendrocytes, the myelin-making cells in the brain and spinal cord, are also killed by the injury and may need to be replaced as part of effective spinal cord repair.

**Neurons:** During the injury, the maximum movement of tissue occurs in the center of the spinal cord; in fact, the spinal cord's core, the gray matter where nerve cell bodies reside, is most susceptible to injury. In addition to neurons mechanically destroyed by the trauma, secondary damage continues to kill neurons, along with other types of cells in the cord, long after the initial injury. Moreover, some neurons die—or at least shrink—later on because their axon has been cut off, usually if the cut has occurred close to the cell body; a spinal cord neuron whose axon was cut or damaged near its terminal end will usually survive.

## Sources for the Appendix

Kalat, J. W. 1995. *Biological Psychology,* 5th ed. Pacific Grove, Calif.: Brooks/Cole Publishing.

McDonald, J. W., and the Research Consortium of the Christopher Reeve Paralysis Foundation. 1999. Repairing the damaged spinal cord. *Scientific American* 281(September):64-73.

Ornstein, R., and R. F. Thompson. 1984. *The Amazing Brain.* Boston: Houghton Mifflin.

Snell, R. S. 1977. *Clinical Neuroanatomy.* Philadelphia: Lippincott-Raven Publishers.

Thompson, R. F. 1993. *The Brain: A Neuroscience Primer.* New York: W. H. Freeman.

Tortora, G. J., and S. Reynolds Grabowski. 1993. *Principles of Anatomy and Physiology.* New York: HarperCollins College Publishers.

Young, W. Personal communication, December 2000.

# NOTES AND SOURCES

## GENERAL

Several resources, listed below, proved indispensable in learning about spinal cord injuries and the structure and function of the nervous system. (Most libraries carry *Scientific American*; the books were still in print in 2001.)

Kalat, J. W. 2001. *Biological Psychology*, 7th ed. Pacific Grove, Calif.: Brooks/Cole Publishing.
A lively, accessible textbook on the brain and other parts of the nervous system.

McDonald, J. W., and the Research Consortium of the Christopher Reeve Paralysis Foundation. 1999. Repairing the damaged spinal cord. *Scientific American* 281(September):64-73.
A concise but comprehensive overview of spinal cord injury and repair strategies.

Ornstein, R., and R. F. Thompson. 1984. *The Amazing Brain*. Boston: Houghton Mifflin.
A popular book about the brain, with gorgeous illustrations by David Macaulay.

Thompson, R. F. 1993. *The Brain: A Neuroscience Primer*, 2d ed. New York: W. H. Freeman.
An excellent introduction to the brain and basic concepts of neuroscience for people with basic knowledge of biology.

# INTRODUCTION

p. xi   **Egyptian papyrus**: The Edwin Smith papyrus, now kept in the Rare Book Room of the New York Academy of Medicine, is named after the American collector who bought it. The papyrus, once thought to be more ancient, is now believed to date to approximately 1550 B.C.E. (J. F. Nunn, *Ancient Egyptian Medicine* [Norman: University of Oklahoma Press, 1996], p. 26.) In addition to descriptions of spinal cord injuries, the papyrus also contains the first reference to the brain anywhere in human records. In describing 48 surgical cases, the author refers to some as "a disease which I shall treat," to other, more serious ones as "a disease which I shall fight," while he puts spinal cord injuries into the category of "a disease that cannot be treated." These phrases have entered numerous textbooks in English and other languages from the German translation of the papyrus by W. Westendorf. In an English translation direct from the ancient Egyptian, they are rendered "an ailment which I will treat," "an ailment with which I will contend," and "an ailment not to be treated" (J. H. Breasted, *The Edwin Smith Surgical Papyrus* [Chicago: University of Chicago Press, 1930]).

p. xii **2 million worldwide:** According to the web site of the International Campaign for Cures of Spinal Cord Injury Paralysis, at *www.campaignforcure.org.* The site provides statistics for several countries and estimates for the number and cost of injuries worldwide. For more U.S. statistics, see the web sites of the National Spinal Cord Injury Statistical Center, *www.spinalcord.uab.edu*, and the National Spinal Cord Injury Association, *www.spinalcord.org.*

p. xiv **50 million Americans, $400 billion:** Brain Facts. Washington, D.C.: Society for Neuroscience, 1997.

# PROLOGUE

p. 3 **Christopher Reeve:** Rosenblatt, R. 1996. New hopes, new dreams. *Time* 148(August 26):28-40; Reeve, C. *Still Me.* 1998. New York: Random House.

p. 4 **Turnbull writes in her autobiography:** Turnbull, B. 1997. *Looking in the Mirror.* Toronto: The Toronto Star.

p. 7 **Lord Nelson was spinally injured:** Beatty, W. 1985. *The Death of Lord Nelson.* London: The Athenaeum.

p. 7 **during World War I:** Ludwig Guttmann. 1973. *Spinal Cord Injuries: Comprehensive Management and Research.* Oxford: Blackwell Scientific Publications, p. 5.

p. 7 **A 1936 surgery textbook:** Quoted in Vladimir Benes. 1968. *Spinal Cord Injury.* London: Bailliere, Tindall & Cassell, p. 6.

p. 8 **headed the famous unit:** Frankel, H. 1999. 100 years after his birth Guttmann's message lives on. *Spinal Cord* 37:461-462.

p. 8 **recalls an observer:** Professor Patrick Wall

p. 8 **"To transform a hopeless":** Susan Goodman. 1986. *Spirit of Stoke Mandeville: The Story of Sir Ludwig Guttmann.* London: Collins, p. 165.

p. 8 **"traveling salesman":** Gingras, G. 1977. *Feet Was I to the Lame.* London: Souvenir Press (E&A).

p. 8 **By 1980:** Young, W. 2000. Molecular and cellular mechanisms of spinal cord injury therapies. P. 249 in *Neurobiology of Spi-*

*nal Cord Injury,* R. G. Kalb and S. M. Strittmatter, eds. Totowa, N.J.: Humana Press.

p. 10 **"I went to Russia for hope":** Waldrep, K., and S. M. Malone. 1996. *Fourth and Long: The Kent Waldrep Story.* New York: Crossroad Publishing, p. 152.

p. 10 **Rick Hansen:** Hansen, R., and J. Taylor. 1999. *Rick Hansen, Man in Motion,* 2d ed. Vancouver: Douglas & McIntyre.

# CHAPTER 1

For **Cajal:**

Cannon, D. 1949. *Explorer of the Human Brain.* New York: Henry Schuman. A biography with a memoir by Sir Charles Sherrington.

DeFelipe, J. 1998. Santiago Ramón y Cajal. Pp. 98-99 in *The MIT Encyclopedia of the Cognitive Sciences,* R. A. Wilson and F. Keil, eds. Cambridge, Mass.: MIT Press.

DeFelipe, J., and E. G. Jones, eds. 1991. *Cajal's Degeneration and Regeneration of the Nervous System.* New York: Oxford University Press.

Ramón y Cajal, S. 1996. *Recollections of My Life.* Cambridge, Mass.: MIT Press. Originally published in English as Volume 8 of *Memoirs of the American Philosophical Society,* 1937.

Sherrington, C. 1935. Santiago Ramón y Cajal, 1852-1934. *Obituary Notices of Fellows of the Royal Society* 4:425-441.

Wasson, T., ed. 1987. *Nobel Prize Winners.* New York: H. W. Wilson, pp. 852-855.

Williams, H. 1954. *Don Quixote of the Microscope.* London: Jonathan Cape.

Cajal's quotes in this chapter are from *Recollections of My Life* and from *Cajal's Degeneration and Regeneration of the Nervous System.*

## For **Tello:**

Collazo Rodríguez, A. F. 1980. Ph.D. dissertation. *Vida y obra de Jorge Francisco Tello.* Universidad Complutense de Madrid, Facultad de Medicina.

de Castro, F. 1981. Cajal y la Escuela Neurológica Española. Madrid: Editorial de la Universidad Complutense.

Martínez Pérez, R. D. 1959. Jorge Francisco Tello en la cátedra. *Revista IBYS* 12(March-April):166-173.

Ramón y Cajal, S. 1923. Discurso leído en la Real Academia Nacional de Medicina en su recepción pública por el académico electo Dr. J. Francisco Tello y contestación por el Excmo. Sr. D. Santiago Ramón y Cajal.

Ricoy, J. 1980. In memoriam: J. F. Tello. *Archivos de Neurobiología* 43:347-352.

Rodríguez-Pérez, A. P. 1959. J. F. Tello y su labor científica. *Trabajos del Instituto Cajal.* Madrid: CSIC.

Tello, J. F. 1952. Recuerdos de Cajal. *Revista IBYS* 3(May-June):3-14.

## For **Golgi:**

Bentivoglio, M. 1999. Life and Discoveries of Camillo Golgi. Nobel Foundation web site, *www.nobel.se.*

Coulston Gillispie, C., ed. 1975. *Dictionary of Scientific Biography.* New York: Charles Scribner's Sons, pp. 459-461.

Mazzarello, P. 1999. *The Hidden Structure.* Oxford: Oxford University Press.

Wasson, T., ed. 1987. *Nobel Prize Winners.* New York: H. W. Wilson, pp. 392-394.

p. 18   **Hero worship in science:** Jacobson, M. 1993. *Foundations of Neuroscience.* New York: Plenum Press, pp. 229-290.

p. 19   **"the gains in brain":** Bloom, F. E. 1975. The gains in brain are mainly in the stain. P. 211 in *The Neurosciences: Paths of Discovery,* F. G. Worden, J. P. Swazey, and G. Adelman, eds. Cambridge, Mass.: MIT Press. According to Dr. Bloom, the originator of the phrase was Dr. Stanley Yolles.

p. 20　**a human hair split 20,000 times:** 20 nanometers. Thompson, R. F. 1993. *The Brain: A Neuroscience Primer.* New York: W. H. Freeman, p. 35.

p. 20　**"Possibly I did":** Quoted in Swazey, J., and F. Worden. 1975. On the nature of research in neuroscience. P. 570 in *The Neurosciences: Paths of Discovery,* F. G. Worden, J. P. Swazey, and G. Adelman, eds. Cambridge, Mass.: MIT Press.

p. 20　**"a postulate necessary":** Golgi, C. 1967. The neuron doctrine—theory and facts. P. 202 in *Nobel Lectures 1901-1921.* Amsterdam: Elsevier Publishing.

p. 21　**The human optic nerve:** Thompson, R. F. 1993. *The Brain: A Neuroscience Primer.* New York: W. H. Freeman, p. 30.

p. 22　**had been awarded the Nobel Prize:** The concept of a basic unit is crucial for the development of a scientific field. Interestingly, the scientist who discovered a basic unit in another discipline was honored at the same time as Golgi and Cajal. The 1906 Nobel Prize in physics went to Joseph John Thompson, the discoverer of the electron.

p. 22　**according to one account:** Mazzarello, P. 1999. *The Hidden Structure.* Oxford: Oxford University Press, p. 312.

p. 22　**one of his biographers:** Paolo Mazzarello

p. 22　**"generally recognized":** Golgi, C. 1967. The neuron doctrine—theory and facts. P. 189 in *Nobel Lectures 1901-1921.* Amsterdam: Elsevier Publishing.

p. 23　**theatrically re-created:** Haines, D. E. 1996. The Cajal-Golgi Nobel presentations of 1906, revisited in 1985; and Culberson, J. L., and D. O. Overman. Golgi, Cajal and the 1906 Nobel Prize. *Proceedings of the Cajal Club,* Vol. 4.

p. 24　**best equipment:** DeFelipe, J., and E. G. Jones. 1992. Santiago Ramón y Cajal and methods in neurohistology. *Trends in Neurosciences* 15:237-246.

p. 24　**"most embarrassing":** Tello F. 1907. La régénération dans les voies optiques. *Travaux du Laboratoire de Recherches Biologiques* 5:237-248.

p. 24　**a peculiar paradox:** Jones, E. G. 1999. Golgi, Cajal and the neuron doctrine. *Journal of the History of the Neurosciences* 8:170-178.

p. 25   **Debates over the neuron theory:** Opposition to the neuron theory was at times exceptionally strong. At a 1937 scientific congress in Königsberg, a speaker announced that the association of German anatomists intended to "bury the neuron theory," according to a memoir of the Hungarian neuroscientist John Szentágothai, who said he was the only person to defend the theory on that occasion. Szentágothai also noted that objections to the theory were particularly strong in Europe and less so in the United States and USSR. Worden, F. G., J. P. Swazey, and G. Adelman, eds. 1975. *The Neurosciences: Paths of Discovery* Cambridge, Mass.: MIT Press, p. 105. See also Jones, E. G. 1994. The neuron doctrine 1891. *Journal of the History of the Neurosciences* 3:3-20.

p. 25   **one last pocket:** Professor Constantino Sotelo, personal communication, April 2000.

p. 28   **once he even failed:** Dr. Francisco J. Martínez-Tello, personal communication, April 2000.

p. 29   **a speaker at a 1980 meeting:** Solsona, F. 1980. Las razones del homenaje del Ateneo a J. F. Tello. P. 8 in Jorge Francisco Tello. Sesión homenaje en el centenario de su nacimiento. Paper read at Ateneo de Zaragoza, June 13, 1980.

p. 29   **in a memorial lecture:** Tello, J. F. 1991. Cajal y su labor histológica. P. 120 in J. DeFelipe and E. G. Jones, eds. *Cajal's Degeneration and Regeneration of the Nervous System.* New York: Oxford University Press.

p. 29   **"a change in regenerative acts":** Tello, F. 1911. La influencia del neurotropismo en la regeneración de los centros nerviosos. *Trabajos del laboratorio de Investigaciones Biológicas* 9:123-159.

# CHAPTER 2

This chapter is based on interviews with people who knew Dr. Windle and on the archival collection of Dr. Windle's papers and films in the History & Special Collections Division, Louise M. Darling Biomedical Library, UCLA. Manuscript Collection No. 112, processed by Pat L. Walter.

p. 32   **humorous aphorism:** Gerard, R. W. 1955. A perspective evaluation of CNS regeneration in mammals. P. 229 in *Regeneration in the Central Nervous System,* W. F. Windle, ed. Springfield, Ill.: Charles C. Thomas.

p. 33   **a scientist who worked in his laboratory:** Dr. Jerald Bernstein

p. 33   **his graduate studies:** Interestingly, Windle was first introduced to spinal cord regeneration in the mid-1920s while doing his graduate studies. Windle was asked to help with a small regeneration study conducted at the University of Chicago by Theodore Koppanyi (whose earlier research in Europe had produced one of the clearest examples of central nervous regeneration in fish) and Ralph Gerard. The two scientists wanted to check whether the fetal mammalian spinal cord could regenerate as well as the fish spinal cord and cut the spines of rat fetuses while in the womb. (Gerard, R. W., and T. Koppanyi. 1926. Studies on spinal cord regeneration in the rat. *American Journal of Physiology* 76:211-212.) The project produced no conclusive evidence and Windle found the experience uninspiring.

p. 36   **first released in 1950:** Windle, W. F., and W. W. Chambers. 1950. Regeneration in the spinal cord of the cat and dog. *Journal of Comparative Neurology* 93:241-258.

p. 36   **told a newspaper reporter:** Furman, B. 1952. New drug may seal a cut spinal cord. *New York Times,* May 20, p. 27.

p. 38   **Levi-Montalcini:** Levi-Montalcini, R. 1955. Neuronal regeneration in vitro. Pp. 54-65 in *Regeneration in the Central Nervous System,* W. F. Windle, ed. Springfield, Ill.: Charles C. Thomas.

p. 39   **a project he carried out in 1940:** Sugar, O., and R. W. Gerard. 1940. Spinal cord regeneration in the rat. *Journal of Neurophysiology* 3:1-19.

p. 39   **He proclaimed that:** Gerard, R. W. 1955 A perspective evaluation of CNS regeneration in mammals. Pp. 231-232 in *Regeneration in the Central Nervous System,* W. F. Windle, ed. Springfield, Ill.: Charles C. Thomas.

p. 40   **"official thaw":** Maddox, S. 1993. *The Quest for Cure.* Washington, D.C.: Paralyzed Veterans of America, p. 9.

p. 40    **"Few of those who attended":** ibid, p. 12.

p. 41    **an all-time low:** Surgeon reports success in treating paraplegia, but results stir debate. 1967. *JAMA* 202:30-31; Veraa, R. 1972. New hope for a paraplegia cure. *Occupational Health Nursing* 20(January):9-11.

p. 42    **"the only one there":** Richard Veraa, quoted in Maddox, S. 1993. *The Quest for Cure.* Washington, D.C.: Paralyzed Veterans of America, p. 15.

p. 43    **the surprise of his life:** The account of Windle's trip to the USSR is based on his October 1974 report to the Director of the National Institute of Neurological Diseases and Stroke.

p. 44    **"meat tenderizer" approach:** Maddox, S. 1993. *The Quest for Cure.* Washington, D.C.: Paralyzed Veterans of America, p. 82.

p. 45    **all rats remained paraplegic:** Guth, L., E. Alburquerque, S. S. Desphande, C. P. Barrett, and E. J. Donati. 1980. Ineffectiveness of enzyme therapy on regeneration in the transected spinal cord of the rat. *Journal of Neurosurgery* 52:73.

p. 45    **proved long-lasting:** Grafstein, B. 2000. Half a century of regeneration research. Pp. 1-18 in *Regeneration in the Central Nervous System,* N. A. Ingoglia and M. Murray, eds. New York: Marcel Dekker.

p. 45    **a model for other such meetings:** In the United States, the Veterans Administration established the Office of Regeneration Research Programs, which started holding annual regeneration conferences in 1985 at the Asilomar Conference Center in Pacific Grove, California, which continue to this day.

# CHAPTER 3

p. 48    **"plasticity":** The term *plasticity*, which covers a broad range of changes and adaptations, has been used to describe different phenomena in the nervous system since the early twentieth century. For a review of the use of the term in neuroscience, see Jones, E. G. 1999. Neurowords: Plasticity and neuroplasticity. *Journal of the History of the Neurosciences* 8:1-3.

p. 50   **the support of others:** "The 'others' who supported my work were, in Oxford, Dr. Max Cowan, Sir Wilfrid Le Gros Clark, and Professor Geoffrey Harris; and in the United States, Drs. Walle Nauta, Irving Diamond, and Ford Ebner," says Dr. Raisman. "It would have been impossible to survive without this support. The ones who opposed were virtually everyone else."

p. 50   **published in *Brain Research*:** Raisman, G. 1969. Neuronal plasticity in the septal nuclei of the adult rat. *Brain Research* 14:25-48.

p. 51   **when she first began to study:** Grafstein, B. 2000. Half a century of regeneration research. Pp. 1-18 in *Regeneration in the Central Nervous System,* N. A. Ingoglia and M. Murray, eds. New York: Marcel Dekker.

p. 51   **Grafstein and her colleagues:** McEwen, B. S., and B. Grafstein. 1968. Fast and slow components in axonal transport of protein. *Journal of Cell Biology* 38:494-508.

p. 53   **conducted numerous studies:** For a scientific review, see Goldberger, M. M, M. Murray, and A. Tessler. 1993. Sprouting and regeneration in the spinal cord; their roles in recovery of function after spinal injury. Pp. 241-264 in *Neuroregeneration*, A. Gorio, ed. New York: Raven Press.

p. 54   **"the gains in brain":** See note for p. 19 above.

# CHAPTER 4

p. 56   **first published in 1980:** Richardson, P. M., U. M. McGuinness, and A. J. Aguayo. 1980. Axons from CNS neurones regenerate into PNS grafts. *Nature* 284:264-265.

p. 56   **"stood the regeneration community":** Dr. Jean Wrathall

p. 56   **"launched a new era":** Dr. Mary Bunge. Many other scientists in the regeneration field share this opinion. For example, Dr. James Fawcett wrote in a 1998 review: "The experiments that really set the field of spinal cord repair into motion in the 1980s were done by Aguayo and his colleagues, and were designed to see whether CNS axons have the ability to regenerate if they are given a permissive environment." Fawcett, J. W. 1998. Review. Spinal cord repair: from experimental models to human application. *Spinal Cord* 36:811-817.

p. 58 **surgeon argued in a 1905 scientific paper:** David Alexander Shirres. Quoted in The Development of Neurology at McGill, a McGill University brochure, by Preston Robb.

p. 59 **proclaimed that the Spaniards had made a mistake:** Le Gros Clark, W. E. 1942, 1943. The problem of neuronal regeneration in the central nervous system. Parts 1 and 2. *Journal of Anatomy* 77:20, 77:251; Le Gros Clark, W. E. 1968. *Chant of Pleasant Exploration: An Autobiography.* Edinburgh: E. & S. Livingstone, pp. 145-147.

p. 60 **coedited an important book:** Kao, C., R. Bunge, and P. Reier, eds. 1983. *Spinal Cord Reconstruction.* New York: Raven Press.

p. 65 **published in 1981:** David, S., and A. J. Aguayo. 1981. Axonal elongation into peripheral nervous system "bridges" after central nervous system injury in adult rats. *Science* 214:931-933.

# CHAPTER 5

p. 68 **first reported in 1987:** Vidal Sanz, M., G. M. Bray, M. P. Villegas-Perez, and A. J. Aguayo. 1987. Axonal regeneration and synapse formation in the superior colliculus by retinal ganglion cells in the adult rat. *Journal of Neuroscience* 7:2894-2907.

p. 69 **legendary Buddhist monk:** the fifth-century Indian monk Bodhidharma, known in Japan as Bodai Daruma

p. 70 **to be published in *Science*:** Keirstead, S. A., M. Rasminsky, Y. Fukuda, D. A. Carter, A. J. Aguayo, and M. Vidal Sanz. 1989. Electrophysiologic responses in hamster superior colliculus evoked by regenerating retinal axons. *Science* 246:255-257.

p. 72 **Yutaka Fukuda:** Sasaki, H., T. Inoue, H. Iso, Y. Fukuda, and Y. Hayashi. 1993. Light-dark discrimination after sciatic nerve transplantation to the sectioned optic nerve in adult hamsters. *Vision Research* 33:877-880.

p. 72 **Kwok-Fai So:** Schneider, G. E., S.-W. You, K.-F. So, D. A. Carter, F. Khan, R. Ellis-Behnkel, and A. Okobi. 2000. Visual function due to regeneration of optic nerve or optic tract through peripheral nerve homografts. *Society for Neuroscience Abstracts* 26(I):611.

p. 72 **Manuel Vidal Sanz:** Avilés-Trigueros, M., Y. Sauvé, R. D. Lund, and M. Vidal Sanz. 2000. Selective innervation of retinorecipient brainstem nuclei by retinal ganglion cell axons regenerating through peripheral nerve grafts in adult rats. *Journal of Neuroscience* 20:361-374.

# CHAPTER 6

p. 74 For **Schwann:**
Coulston Gillispie, C., ed. 1975. *Dictionary of Scientific Biography.* New York: Charles Scribner's Sons, pp. 240-245.
Haymaker, W., and F. Schiller, eds. 1970. *The Founders of Neurology.* Springfield, Ill.: Charles C. Thomas, pp. 77-80.
Porter, R., ed. 1994. *The Biographical Dictionary of Scientists,* 2nd ed. New York: Oxford University Press, pp. 610-611.

p. 76 **would do a better job:** Crucial evidence for the important role that Schwann cells, rather than other elements in peripheral nerve tissue, play in regeneration was supplied by studies by Professor Martin Berry of King's College, London. Berry, M., L. Rees, S. Hall, P. Yiu, and J. Sievers. 1988. Optic axons regenerate into sciatic nerve isografts only in the presence of Schwann cells. *Brain Research Bulletin* 20:223-231.

p. 77 **1 to 10 meters per second:** Kalat, J. W. 1995. *Biological Psychology,* 5th ed. Pacific Grove, Calif.: Brooks/Cole Publishing, p. 50.

p. 77 **one neuroscience professor:** Dr. John Steeves

p. 77 **about 100 meters a second:** Thompson, R. F. 1993. *The Brain.* New York: W. H. Freeman, p. 59.

p. 77 **Human babies:** For more on the development of the

nervous system in fetuses and babies, see Eliot, L. 1999. *What's Going on in There?* New York: Bantam Books.

p. 79 **name was inspired:** according to Dr. Barth Green, the neurosurgeon who helped found the Miami Project. Much of the Project's high profile has stemmed from the involvement of the Buoniconti family, which created a fund for spinal cord research after Marc Buoniconti, son of ex-Miami Dolphin football star Nick, broke his neck playing college football in 1985.

p. 80 **one of the team leaders:** Dr. Gerald Fischbach

p. 81 **Schwann cell cables have revealed:** For a scientific review, see Bunge, M. B., and N. Kleitman. 1999. Neurotrophins and neuroprotection improve axonal regeneration into Schwann cell transplants placed in transected adult rat spinal cord. Pp. 631-645 in *CNS Regeneration,* M. H. Tuszynski and J. Kordower, eds. San Diego, Calif.: Academic Press.

## CHAPTER 7

p. 83 **had been raised earlier:** For example, as early as 1982, Professor Martin Berry of King's College London published a scientific paper outlining a hypothesis that myelin debris may be responsible for the failure of the mammalian central nervous system to regenerate. Berry, M. 1982. Post-injury myelin-breakdown products inhibit axonal growth: an hypothesis to explain the failure of axonal regeneration in the mammalian central nervous system. *Bibliotheca Anatomica* 23:1-11.

p. 84 **chief executive:** Peter Banyard

p. 86 **managed to produce an antibody:** The antibody was identified by Dr. Pico Caroni, a biochemist and cell biologist who now heads his own laboratory at the Friedrich Miescher Institute in Basel, and Dr. Paolo Paganetti, now head of a CNS research team at Novartis, Basel.

p. 90 **reported in *Nature:*** Schnell, L., and M. Schwab. 1990. Axonal regeneration in the rat spinal cord produced by an antibody against myelin-associated neurite growth inhibitors. *Nature* 343:269-272.

p. 92 **reported in *Nature*:** Bregman, B.S., E. Kunkel-Bagden, L. Schnell, H. N. Dai, D. Gao, and M. Schwab. 1995. Recovery from spinal cord injury mediated by antibodies to neurite growth inhibitors. *Nature* 378:498-501; editorial by Clifford Woolf. Overcoming inhibition. 378:439-440.

## CHAPTER 8

p. 96 **two independent research teams:** Mukhopadhyay, G., P. Doherty, F. S. Walsh, P. R. Crocker, and M. T. Filbin. 1994. A novel role for myelin-associated glycoprotein as an inhibitor of axonal regeneration. *Neuron* 13:757-767. McKerracher, L., S. David, D. L. Jackson, V. Kottis, R. J. Dunn, and P. E. Braun. Identification of myelin-associated glycoprotein as a major myelin-derived inhibitor of neurite growth. *Neuron* 13:805-811.

p. 98 **finally purified the protein:** Spillmann, A. A., C. E. Bandtlow, F. Lottspeich, F. Keller, and M. E. Schwab. Identification and characterization of a bovine neurite growth inhibitor (bNI-220). *Journal of Biological Chemistry* 273:19283-19293.

p. 101 **scientific ammunition:** Davies, S. J., M. T. Fitch, S. P. Memberg, A. K. Hall, G. Raisman, and J. Silver. 1997. Regeneration of adult axons in white matter tracts of the central nervous system. *Nature* 390:680-683. Davies, S. J., D. R. Goucher, C. Doller, and J. Silver. 1999. Robust regeneration of adult sensory axons in degenerating white matter of the adult rat spinal cord. *Journal of Neuroscience* 19:5810-5822.

p. 102 **a treatment that temporarily strips:** Keirsted, H. S., S. J. Hasan, G. D. Muir, and J. D. Steeves. 1992. Suppression of the onset of myelination extends the permissive period for the functional repair of embryonic spinal cord. *Proceedings of the National Academy of Sciences of the U.S.A.* 89:11664-11668.

p. 102 **a therapeutic vaccine:** Huang, D. W., L. McKerracher, P. E. Braun, and S. David. 1999. A therapeutic vaccine approach to stimulate axon regeneration in the adult mammalian spinal cord. *Neuron* 24:639-647.

p. 102 **at least three teams:** Lehmann, M., A. E. Fournier, I.

Selles-Navarro, P. Dergham, N. Leclerc, G. Tigyi, and L. McKerracher. 1999. Inactivation of the small GTP-binding protein Rho promotes CNS axon regeneration. *Journal of Neuroscience* 19:7537-7547. Cai, D., Y. Shen, M. E. De Bellard, S. Tang, and M. T. Filbin. 1999. Prior Exposure to neurotrophins blocks inhibition of axonal regeneration by MAG and myelin via a cAMP-dependent mechanism. *Neuron* 22:89-101. Benowitz, L. I., D. E. Goldberg, J. R. Madsen, D. Soni, and N. Irwin. 1999. Inosine stimulates extensive axon collateral growth in the rat corticospinal tract after injury. *Proceedings of the National Academy of Sciences of the U.S.A.* 96:13486-13490.

## CHAPTER 9

p. 104 **"the enlightenment":** information brochures of the Karolinska Institute.

p. 105 **editorial:** Young, W. 1996. Spinal cord regeneration. *Science* 273:451.

p. 105 *Time* **magazine wrote:** Rosenblatt, R. 1996. New hopes, new dreams. *Time* 148(August 26):28-40.

p. 105 **"a throb of excitement":** Susan Howley, director of research at the Christopher Reeve Paralysis Foundation

p. 106 **strong tradition:** Three eminent Swedish professors at the University of Göteborg made groundbreaking contributions to the study of Parkinson's disease. Nobel laureate Arvid Carlsson provided insights into the molecular basis of Parkinson's. Nils-Åke Hillarp, who later moved to the Karolinska Institute (where he created an enthusiastic following of young scientists, including Olson), together with Bengt Falck, who moved to the University of Lund, created the first method for tracing the class of neurotransmitters that includes dopamine. The 1960s Falck–Hillarp method is still the only technique that makes it possible to visualize any kind of neurotransmitter directly, rather than by indirect means such as using attached antibodies.

p. 106 *Newsweek* **article:** Begley, S. 1986. From hearts to minds. *Newsweek* 107(April 14):62-63.

p. 112 **one veteran regeneration researcher:** Professor Raymond Lund

p. 113 **"the happiest moment":** Goldberg, J. 1997. Mending spinal cords. *Discover* 18(January):84-85.

# CHAPTER 10

p. 120 **methylprednisolone:** A trial of some 500 patients conducted at 10 medical centers in the United States showed that the steroid hormone called methylprednisolone, when given in extremely high doses within eight hours of spinal cord injury, helped reduce disability after the trauma. (Bracken, M. B., M. J. Shepard, W. F. Collins, T. R. Holford, W. Young, D. S. Baskin, H. M. Eisenberg, E. Flamm, L. Leo-Summers, J. Maroon, L. F. Marshall, P. L. Perot, J. Piepmeier, V. K. H. Sonntag, F. C. Wagner, J. E. Wilberger, and H. R. Winn. 1990. A randomized, controlled trial of methylprednisolone or naloxone in the treatment of acute spinal-cord injury. *New England Journal of Medicine* 322:1405-1411.) On March 30, 1990, the U.S. National Institutes of Health took the unusual step of announcing the trial's results before they were published in a scientific journal so that emergency room physicians could start giving the therapy to spinally injured patients as soon as possible. The rather inexpensive steroid, commonly available in hospitals, had been used for years to treat shock and prevent swelling of the brain from injury or stroke. In spinal cord trauma, however, it works only if given at a dose 100 times larger than usual, which suggests that its normal anti-inflammatory effect alone cannot account for its action; other mechanisms, not fully explained, are believed to be involved. In the United States the use of methylprednisolone after spinal cord injury is now standard practice. However, the drug is unpopular in some countries, such as the United Kingdom, and some American physicians question its risk:benefit ratio (although most would probably never dare withhold it, for fear of being sued). (See, for example, Nesathurai, S. 1998. Steroids and spinal cord injury: revisiting the NASCIS 2 and NASCIS 3 trials. *Journal of Trauma* 45:1088-1093.)

p. 120 **Sygen had indeed:** Geisler, F. H., F. C. Dorsey, and W. P. Coleman. 1991. Recovery of motor function after spinal-cord injury—a randomized, placebo-controlled trial with GM-1 ganglioside. *New England Journal of Medicine* 325:1659-1660.

p. 121 **wrote *Time*:** Rosenblatt, R. 1996. New hopes, new dreams. *Time* 148(August 26):28-40.

p. 122 **director of research at the Christopher Reeve Paralysis Foundation:** Susan Howley

p. 122 **In his essay:** Krauthammer, C. 2000. Restoration, reality and Christopher Reeve. *Time* 155(February 14):76.

p. 123 **Byrne wrote a column:** Byrne, D. 2000. Reeve Takes Stand for Hope. *Chicago Sun-Times* February 6, p. 21A.

p. 123 **a follow-up column:** Byrne, D. 2000. Don't Tread on Our Hope. *Chicago Sun-Times* February 20, p. 38A.

p. 126 **$10 billion per year:** Spinal Cord Injury: Emerging Concepts. Proceedings of an NIH Workshop, September 30-October 1, 1996.

p. 126 **close to $550,000:** National Spinal Cord Injury Statistical Center, at *www.spinalcord.uab.edu*

p. 126 **It costs on average $500 million:** according to the Pharmaceutical Research and Manufacturers of America

p. 126 **including federal funding:** In 1999, for instance, direct government investment in spinal cord injury research in the United States was $70 million, according to the International Campaign for Cures of Spinal Cord Injury Paralysis, at *www.campaignforcure.org*

p. 129 **review article:** Horner, P. J., and F. H. Gage. 2000. Regenerating the damaged central nervous system. *Nature* 407:963-970.

p. 131 **recent editorial:** Nash, J., and A. Pini. 1996. Making the connections in nerve regeneration. *Nature Medicine* 2:25-26.

# CHAPTER 11

p. 135 **reported in 1985 in *Science*:** Schwartz, M., M. Belkin, A. Harel, A. Solomon, V. Lavie, M. Hadani, I. Rachailovich, and C.

Stein-Izsak. 1985. Regenerating fish optic nerves and a regeneration-like response in injured optic nerves of adult rabbits. *Science* 228:600-603.

p. 138   **treated rats again showed signs of recovery:** Rapalino, O., O. Lazarov-Spiegler, E. Agranov, G. Velan, E. Yoles, A. Solomon, R. Gepstein, A. Katz, M. Belkin, M. Hadani, and M. Schwartz. 1998. Implantation of stimulated macrophages leads to partial recovery of paraplegic rats. *Nature Medicine* 4:814-821.

p. 141   **Secondary damage:** For a more detailed description of secondary damage and strategies to contain it, see McDonald, J. W., and the Research Consortium of the Christopher Reeve Paralysis Foundation. 1999. Repairing the damaged spinal cord. *Scientific American* 281(September):64-73.

p. 142   **T cells indeed protected:** Moalem, G. R., R. Leibowitz-Amit, E. Yoles, F. Mor, I. Cohen, and M. Schwartz. 1999. Autoimmune T cells protect neurons from secondary degeneration after central nervous system axotomy. *Nature Medicine* 5:49-55; and Hauben, E., O. Butovsky, U. Nevo, E. Yoles, G. Moalem, E. Agranov, F. Mor, R. Leibowitz-Amit, S. Pevsner, S. Akselrod, M. Neeman, I. R. Cohen, and M. Schwartz. Passive or active immunization with myelin basic protein promotes recovery from spinal cord contusion. *Journal of Neuroscience* 20:6421-6430.

p. 142   **Cohen had been saying:** Cohen, I. R. 2000. *Tending Adam's Garden: Evolving the Cognitive Immune Self.* London: Academic Press.

# CHAPTER 12

p. 146   **wrote *Science*:** Breakthrough of the year. 1999. *Science* 286:2238-2239.

p. 147   **toppled the old dogma:** Kempermann, G., and F. Gage. 1999. New nerve cells for the adult brain. *Scientific American* 280(May):48-53.

p. 147   **first animal studies:** McDonald, J. W., X. Z. Liu, Y. Qu, S. Liu, S. K. Mickey, D. Turetsky, D. I. Gottlieb, and D. W. Choi. 1999. Transplanted embryonic stem cells survive, differentiate and promote recovery in injured rat spinal cord. *Nature Medicine* 5:1410-1412; and Liu, S., Y. Qu, T. Stewart, M. Howard, S. Chakrabortty, T. Holekamp, and J. W. McDonald. 2000. Embryonic stem cells differentiate into oligodendrocytes and myelinate in culture and after spinal cord transplantation. *Proceedings of the National Academy of Sciences of the U.S.A.* 97:6126-6131.

p. 148   **In the brain:** The first transplants of nerve tissue into the brain were performed in 1890, with poor results. In the early 1970s, the field of brain transplantation was revived by a series of successful studies. For a historical review, see Björklund, A. 1991. Neural transplantation—an experimental tool with clinical possibilities. *Trends in Neurosciences* 14:319-322; and Björklund, A., and U. Stenevi. 1985. Intracerebral neural grafting: a historical perspective. Pp. 3-14 in *Neural Grafting in the Mammalian CNS,* A. Björklund and U. Stenevi, eds. Amsterdam: Elsevier Science Publishers.

p. 149   **"by far the most difficult":** Das, G. 1983. Neural transplantation in the spinal cord of the acute mammal. P. 391 in *Spinal Cord Reconstruction,* C. Kao, R. Bunge, and P. Reier, eds. New York: Raven Press.

p. 150   **A Nose for Treatment:** For scientific reviews, see Franklin, R. J. M., and S. C. Barnett. 1997. Do olfactory glia have advantages over Schwann cells for CNS repair? *Journal of Neuroscience Research* 50:1-8; and Franklin, R. J. M., and S. C. Barnett. Olfactory ensheathing cells and CNS regeneration—the sweet smell of success? *Neuron* 28:1-4.

p. 151   **reported in 2000:** Ramón-Cueto, A., I. Cordero, F. Santos-Benito, and J. Avila. 2000. Functional recovery of paraplegic rats and motor axon regeneration in their spinal cords by olfactory ensheathing glia. *Neuron* 25:425-435.

p. 152   ***New York Times:*** Grady, D. 1997. Spine researchers seek recipe for regeneration. September 30, p. F1.

p. 152   **nose glia results appeared:** Li, Y., P. M. Field, and G. Raisman. 1997. Repair of adult rat corticospinal tract by transplants of olfactory ensheathing cells. *Science* 277:2000-2002.

# CHAPTER 13

p. 153 **"special food"**: DeFelipe, J., and E. G. Jones, eds. 1991. *Cajal's Degeneration and Regeneration of the Nervous System.* New York: Oxford University Press, p. 749.

p. 153 **"agent"**: Levi-Montalcini, R. 1955. Neuronal regeneration in vitro. Pp. 54-65 in *Regeneration in the Central Nervous System,* W. F. Windle, ed. Springfield, Ill.: Charles C. Thomas.

p. 154 For **Levi-Montalcini**:

Grafstein, B. 2000. Half a century of regeneration research. Pp. 1-18 in *Regeneration in the Central Nervous System,* N. A. Ingoglia and M. Murray, eds. New York: Marcel Dekker.

Levi-Montalcini, R. 1975. NGF: An Uncharted Route. Pp. 244-265 in *The Neurosciences: Paths of Discovery,* F. G. Worden, J. P. Swazey, and G. Adelman G, eds. Cambridge, Mass.: MIT Press.

Levi-Montalcini, R., and Calissano, P. 1979. The nerve-growth factor. *Scientific American* 240(June):44-53.

Levi-Montalcini, R. 1987. The nerve growth factor 35 years later. *Science* 237:1154-1162.

Levi-Montalcini, R. 1988. *In Praise of Imperfection: My Life and Work.* Translated by Luigi Attardi. New York: Basic Books.

Wasson, T., ed. 1987. *Nobel Prize Winners.* New York: H. W. Wilson, pp. 625-627.

Interview: Rita Levi-Montalcini. 1988. *Omni* 10(March):70-72.

The heart & mind of a genius. 1987. *Vogue* 177:480.

NGF may hold the key—but to what? 1977. *Science News* 111(May 21):330-335.

The 1986 Nobel Prize for physiology or medicine. 1986. *Science* 234:543-544.

Nobel Prizes. 1986. *Time* 128(October 27):67.

The Nobel Prizes: Physiology or medicine. 1986. *Scientific American* 255(December):73-74.

Two pioneers in growth of cells win Nobel Prize. 1986. *New York Times,* October 14, p. A1.

p. 156　**one of his former students:** the American anatomist Elmer D. Bueker

p. 158　**according to one hypothesis:** Levi-Montalcini, R. 1987. The nerve growth factor 35 years later. *Science* 237:1154-1162.

p. 159　**the first method for obtaining NGF:** Shooter, E. 2001. Early days of the nerve growth factor proteins. *Annual Review of Neuroscience* 24:601-629.

p. 159　*Science News* **still ran:** Arehart-Treichel, J. 1977. NGF may hold the key—but to what? *Science News* 111(May 21):330-335.

p. 159　**The second member of the NGF family:** Barde, Y.-A., D. Edgar, and H. Thoenen. 1982. Purification of a new neurotrophic factor from mammalian brain. *EMBO Journal* 1:549-553.

p. 160　**250,000 per minute:** Thompson, R. F. 1993. *The Brain: A Neuroscience Primer,* 2d ed. New York: W. H. Freeman, p. 299.

p. 161　**Many neuroscientists believed:** Cowan, W. M. 2001. Viktor Hamburger and Rita Levi-Montalcini: The path to the discovery of nerve growth factor. *Annual Review of Neuroscience* 24:551-600.

p. 161　**"peripheral in every sense of the word":** The phrase belongs to Viktor Hamburger. Quoted in Levi-Montalcini, R. 1987. The nerve growth factor 35 years later. *Science* 237:1154-1162.

p. 162　**the first gene therapy study:** Tuszynski, M. H., D. A. Peterson, J. Ray, A. Baird, Y. Nakahara, and F. H. Gage. 1994. Fibroblasts genetically modified to produce nerve growth factor induce robust neuritic ingrowth after grafting to the spinal cord. *Experimental Neurology* 126:1-14.

p. 163　**thanks to the growth factor alone:** Tuszynski, M. H., R. Grill, and A. Blesch. 1999. Spinal cord regeneration: growth factors, inhibitory factors and gene therapy. Pp. 605-629 in *CNS Regeneration,* M. H. Tuszynski and J. Kordower, eds. San Diego, Calif.: Academic Press.

p. 163　**scientists are trying to elucidate the neuronal preferences:** One of the leading teams in growth-factor gene therapy of the spinal cord, working at the Medical College of Pennsylvania/

Hahnemann University in Philadelphia has tested the role of the growth factor BDNF in spinal cord injury. Interestingly, the neurons that regenerated in this study belonged to a different tract in the spinal cord from the ones involved in the San Diego NGF experiments. Liu, Y., D. Kim, B. T. Himes, S. Y. Chow, T. Schallert, M. Murray, A. Tessler, and I. Fischer. 1999. Transplants of fibroblasts genetically modified to express BDNF promote regeneration of adult rat rubrospinal axons and recovery of forelimb function. *Journal of Neuroscience* 19:4370-4387.

p. 163    **ignore the inhibitors:** Bregman, B. S. 2000. Transplants and neurotrophins modify the response of developing and mature CNS neurons to spinal cord injury. Pp. 169-193 in *Neurobiology of Spinal Cord Injury*, R. G. Kalb and S. M. Strittmatter, eds. Totowa, N.J.: Humana Press.

p. 164    **neutralize the chemical scar:** Berry, M., and A. Logan. 2000. The response of the CNS to penetrant injury: Scarring and regeneration of axons. Pp. 479-498 in *Neurosurgery*, A. Crockard, R. Hayward, and J. Hoff, eds. London: Blackwell Science.

# CHAPTER 14

p. 166    **100 billion; 1,000 to 10,000 connections:** Neuroscience Education: Brain Facts and Figures, at *http://faculty.washington. edu/chudler/ehc.html.*

p. 166    **more than the number of stars:** The number of stars in our galaxy, the Milky Way, is approximately 10 billion—1 followed by 10 zeros.

p. 166    **a competing theory took over:** Hamburger, V. 1991. Foreword. Pp. vii-xi in *Cajal's Degeneration and Regeneration of the Nervous System*, J. DeFelipe and E. G. Jones, eds. New York: Oxford University Press.

p. 167    **challenged the views of his teachers:** Sperry's mentor at the University of Chicago was Paul Weiss, one of the most influential biologists of his time and a prominent advocate of the physical guidance idea. Sperry's complex relationship with his mentor is being explored by Dr. Sabine Brauckmann, who is working on

Weiss's biography. (Brauckmann, S. 1998. From resonance to chemoaffinity: The controversy between Paul A. Weiss and Roger W. Sperry. Talk presented at the Fishbein Center, University of Chicago, November 4.)

p. 168 **some of Sperry's devotees:** Grafstein, B. 2000. Half a century of regeneration research. Pp. 1-18 in *Regeneration in the Central Nervous System,* N. A. Ingoglia and M. Murray, eds. New York: Marcel Dekker.

p. 168 **some seasoned researchers:** Sotelo, C. 1999. From Cajal's chemotaxis to the molecular biology of axon guidance. *Brain Research Bulletin* 50:395-396.

p. 170 **netrin report appeared:** Serafini, T., T. E. Kennedy, M. J. Galko, C. Mirzayan, T. M. Jessell, and M. Tessier-Lavigne. 1994. The netrins define a family of axon outgrowth-promoting proteins homologous to *C. elegans* UNC-6. *Cell* 78:409-424; and Kennedy, T. E., T. Serafini, J. R. de la Torre, and M. Tessier-Lavigne. 1994. Netrins are diffusable chemotropic factors for commisural axons in the embryonic spinal cord. *Cell* 78:425-435.

p. 171 **the title of an editorial:** Goodman, C. 1994. The likeness of being: phylogenetically conserved molecular mechanisms of growth cone guidance. *Cell* 78:353-356.

p. 172 **a 1996 review:** Tessier-Lavigne, M. and C. Goodman. 1996. The molecular biology of axon guidance. *Science* 274:1123-1133.

# CHAPTER 15

Pearson, K. 1976. The control of walking. *Scientific American* 235(December):72-86.

Harkema, S. J., B. H. Dobkin, and V. R. Edgerton. 2000. Pattern generators in locomotion: implications for recovery after spinal cord injury. *Topics in Spinal Cord Injury Rehabilitation* 6(2):82-96.

Rossignol, S. 2000. Locomotion and its recovery after spinal injury. *Current Opinion in Neurobiology* 10:708-716.

p. 179    **"Evolution rarely throws out":** Grillner, S. 1996. Neural networks for vertebrate locomotion. *Scientific American* 274(January):48-53.

p. 180    **"continuing series of surprises":** Wolpaw, J. R. 1997. The complex structure of a simple memory. *Trends in Neurosciences* 20:588-594.

p. 181    **the brain may not be necessary at all:** Grau, J. W., and R. Joynes. 2001. Pavlovian and instrumental conditioning within the spinal cord: Methodological issues. Pp. 12-53 in *Spinal Cord Plasticity: Alterations in Reflex Function,* J. W. Grau, ed. Boston: Kluwer Academic Publishers.

p. 185    **One of Wernig's success stories:** Wickelgren, I. 1998. Teaching the spinal cord to walk. *Science* 279:319-321.

## CHAPTER 16

p. 193    **the first to demonstrate:** Houle, J. D. 1991. Demonstration of the potential for chronically injured neurons to regenerate axons into intraspinal peripheral nerve grafts. *Experimental Neurology* 113:1-9.

p. 194    **the situation is even more complicated:** Houle, J. D., and J.-H. Ye. 1999. Survival of chronically-injured neurons can be prolonged by treatment with neurotrophic factors. *Neuroscience* 94:929-936.

p. 195    **retained their ability:** Grill, R. J., A. Blesch, and M. H. Tuszynski. 1997. Robust growth of chronically injured spinal cord axons induced by grafts of genetically modified NGF-secreting cells. *Experimental Neurology* 148:444-452.

p. 196    **15 years after injury:** Mark Tuszynski, reported at the 2d ISRT Research Network Meeting, London, September 24-26, 1999.

## CHAPTER 17

p. 199    ***Science*headline:** Barinaga, M. 1999. Turning thoughts into actions. *Science* 286:888-890.

p. 200   **"That's a big emotional change":** Chase, V. D. 2000. Mind over muscles. *Technology Review* 103(March/April):39-45.

p. 201   **FES:** FES Information Center in Cleveland provides educational resources on functional electrical stimulation. Contact information: telephone: 800-666-2353; e-mail: *fesinfo@po.cwru.edu*; internet: *http://feswww.fes.cwru.edu.*

p. 202   **Some 2,000 people:** Dr. Graham Creasey, personal communication, October 2000. See also Chae, J., K. Kilgore, R. Triolo, and G. Creasey. 2000. Functional neuromuscular stimulation in spinal cord injury. *Physical Medicine and Rehabilitation Clinics of North America* 11:209-225.

p. 203   **stepping movements in healthy cats:** Mushahwar, V. K., D. F. Collins, and A. Prochazka. 2000. Spinal cord microstimulation generates functional limb movements in chronically implanted cats. *Experimental Neurology* 163:422-429.

# EPILOGUE

p. 205   **a local reporter:** Carnahan, A. 2000. Paralyzed woman regains some movement after pioneering surgery. *Denver Rocky Mountain News,* December 3, p. 12A.

# A Special
# Acknowledgment

Dozens of researchers generously gave me of their time to talk about their own studies and the field of spinal cord repair in general. Some gave interviews lasting several hours, others answered questions at scientific conferences, by e-mail, or over the phone. Some are heads of research institutes, others are students. Here, in alphabetical order, is the list of scientists and physicians engaged in research, as well as members of advocacy groups for spinal cord injury research, whom I interviewed during two years of working on the book. To each of them I express my thanks.

*Researchers:*

| | |
|---|---|
| Eugenia Agranov | Antonio Fernández de Molina |
| Albert Aguayo | Dalton Dietrich |
| Christine Bandtlow | Volker Dietz |
| Susan Barnett | Reggie Edgerton |
| Michael Beattie | James Fawcett |
| Michael Belkin | Michael Fehlings |
| Jerald Bernstein | Marie Filbin |
| Martin Berry | Gerald Fischbach |
| John Bethea | Tamar Flash |
| Anders Björklund | Matt Fraidakis |
| Eran Blaugrund | Hans Frankel |
| Andrew Blight | Yutaka Fukuda |
| Richard Borgens | Valentin Fulga |
| Michael Bracken | Fred Gage |
| Garth Bray | Fred Geisler |
| Barbara Bregman | Harry Goldsmith |
| Jacqueline Bresnahan | Murray Goldstein |
| Mary Bunge | Bernice Grafstein |
| Blair Calancie | James Grau |
| Henrich Cheng | Barth Green |
| Mary Ellen Cheung | Sten Grillner |
| Arlene Chiu | Moshe Hadani |
| Dennis Choi | Edward Hall |
| Carmine Clemente | Adrian Harel |
| Avi Cohen | Ehud Hauben |
| Irun Cohen | William Heetderks |
| Carl Cotman | John Houle |
| Graham Creasey | Andrea Huber |
| James Culberson | Lee Illis |
| Joseph Culotti | Patrick Jacobs |
| Samuel David | Lyn Jakeman |
| Stephen Davies | Edward Jones |
| Vincent DeCrescito | Gong Ju |
| Javier DeFelipe | Carl Kao |
| | Timothy Kennedy |
| | Avihu Klar |

Naomi Kleitman

Michel Kliot

Isabel Klusman

Martin Kuchler

Orly Lazarov-Spiegler

Michel Lévesque

Arkady Livshits

Rodolfo Llinás

Mirit Lotan

Raymond Lund

Jim Lynskey

Pierre Magistretti

Colleen Manitt

Mark Marchionni

Marietta McAtee

Ronald McKay

Lisa McKerracher

Gillian Muir

Marion Murray

Vivian Mushahwar

Shanker Nesathurai

Toomas Neuman

Manuel Nieto-Sampedro

Avi Ohry

Lars Olson

Grigori Orlovsky

Sanford Palay

Alejandra Pazos

Regino Perez-Polo

Hugh Perry

Alexander Rabchevsky

Geoffrey Raisman

Almudena Ramón-Cueto

Michael Rasminsky

William Regelson

Paul Reier

Peter Richardson

Arthur Roach

Douglas Ross

Serge Rossignol

Lisa Schnell

Martin Schwab

Andrew Schwartz

Michal Schwartz

Åke Seiger

Fredrick Seil

Michael Sela

Tito Serafini

Eli Sercarz

Elena Shapkova

Mark Shik

Eric Shooter

Jerry Silver

John Sladek

Kwok-Fai So

Constantino Sotelo

Christian Spenger

Adrian Spillmann

John Steeves

Elke Stein

Bradford Stokes

Charles Tator

Francisco Martínez-Tello

Marc Tessier-Lavigne

Wolfram Tetzlaff

Allan Tobin

Ronald Triolo

Mark Tuszynski

Manuel Vidal Sanz

Michael Walker

Patrick Wall

Anton Wernig

Jonathan Wolpaw          Eti Yoles
Jean Wrathall            Wise Young
Yosef Yarden


*Representatives of spinal injury research advocacy organizations:*

Peter Banyard            Ulrich Schellenberg
John Cavanagh            Barbara Turnbull
Rick Hansen              Kent Waldrep
Susan Howley             Bob Yant
Alan Reich

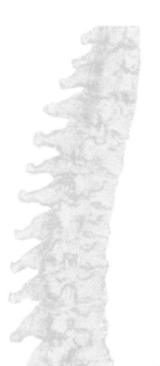

# Photo Credits

*Plate numbers appear in parentheses.*

| | |
|---|---|
| Ramón y Cajal (Pl. 1) | Courtesy of Instituto Cajal, CSIC, Madrid, Spain |
| Francisco Tello (Pl. 1) | Courtesy of Instituto Cajal, CSIC, Madrid, Spain |
| Cajal's drawing of a nerve cell (Pl. 1) | Courtesy of Instituto Cajal, CSIC, Madrid, Spain |
| William Windle (Pl. 2) | Courtesy of the History & Special Collections Division, Louise M. Darling Biomedical Library, UCLA |

| | |
|---|---|
| Melissa Holley (Pl. 2) | Courtesy of Gwen Holley |
| Group photo, NIH conference (Pl. 2) | Courtesy of Neuroscience History Archives, Brain Research Institute, UCLA |
| William Chambers and Chan-nao Liu, with wives (Pl. 3) | Courtesy of Alma Chambers |
| Albert Aguayo (Pl. 3) | Courtesy of the Audiovisual Department, McGill University Health Center |
| Susan Keirstead, Yutaka Fukuda, and Michael Rasminsky (Pl. 3) | Courtesy of Michael Rasminsky |
| Mary and Richard Bunge (Pl. 4) | Courtesy of The Miami Project to Cure Paralysis |
| Martin Schwab with Christopher and Dana Reeve (Pl. 4) | Yana Bridle |
| Henrich Cheng (Pl. 5) | Courtesy of Taipei Veterans General Hospital |
| Lars Olson (Pl. 5) | Stig-Göran Nilsson |
| Treadmill (Laufband) therapy (Pl. 5) | Courtesy of Anton Wernig |
| Rita Levi-Montalcini (Pl. 6) | Courtesy of Rita Levi-Montalcini |
| Sir Ludwig Gutmann (Pl. 6) | From the 1964 portrait by Sir James Gunn |
| Roger Sperry (Pl. 6) | Courtesy of the Archives, California Institute of Technology |
| Wise Young (Pl. 7) | 1998 Rutgers/Nick Romanenko |
| Dennis Byrd at a mike (Pl. 7) | Bill Hickey, Allsport; *Scientific American*, September 1999, 281:64-73 |

Dennis Byrd fallen on the field (Pl. 7)

David Drapkin, NFL Photos; *Scientific American*, September 1999, 281:64-73

Michal Schwartz (Pl. 8)

Gadi Dagon

Post-injury "vaccination" (Pl. 8)

*Journal of Neuroscience* 2000; 20(17):6421-30

Author photo (jacket flap)

Nili Aslan

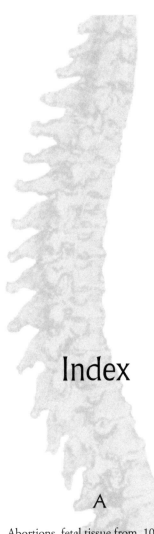

# Index

# T

Tello, Francisco, 24, 28-30, 32, 39, 58, 59, 225
Teratomas, 148
Tessier-Lavigne, Marc, 168-169, 172, 174, 175
Tetraplegia, 3, 200-201
Tetzlaff, Wolfram, 194
Texas A&M University, 181
Thalamus, 214
Thompson, Joseph John, 226
Tissue cultures
  ensheathing glia, 152
  immortal stem cells, 146
  and nerve growth factor, 157-158
  Schwann cells, 79-80
Tracing technologies
  effect on regeneration research, 51-52, 53, 54, 71
  fluorescent, 89
  horseradish enzyme, 62-65, 71
  for peripheral nerve bridges, 55-56, 62-65, 68, 71
  radioactive, 48, 51-52, 53-54, 56
Treadmill therapy, 177, 182-186, 187-188, 190
Triolo, Ronald, 203
Turnbull, Barbara, 4, 141, 223
Tuszynski, Mark, 162, 195

# U

*unc* genes, 171
United States. *See also individuals and research institutions*
  human trials, xiii
University Hospital Balgrist, 187
University of
  Arkansas for Medical Sciences, 192
  Bonn, 184
  British Columbia, 102, 125, 194
  California at Berkeley, 171
  California at Irvine, 91, 113
  California at Los Angeles, 182, 184, 186, 188, 190
  California at San Diego, 162, 174, 195, 196
  California at San Francisco, 168
  Cambridge, 152
  Chicago, 53, 167, 228, 242
  Copenhagen, 181
  Florida, 149, 193
  Glasgow, 152
  Göteborg, 235
  Hong Kong, 72
  Innsbruck, 96
  Lund, 57, 107, 149, 235
  Madrid, San Carlos Faculty of Medicine, 24
  Manitoba, 63
  Maryland, 44, 149
  Miami School of Medicine, 79
  Montreal, 102, 164, 182, 183
  Murcia, 72
  Oxford, 49
  Pennsylvania, 34, 39, 48, 125
  Southhampton, 134
  Toronto, 113, 170
  Turin, 155
  Wisconsin School of Medicine, 74
  Zurich, 84, 85
Urinary tract infections, 7
U.S. Food and Drug Administration, xiii, 5, 6, 7, 101, 121, 128, 129, 139, 148, 198, 201, 202

# V

Vaccines
  neuroprotective, 141-144
  therapeutic, 102
Van Gehuchten, Arthur, 21
VanDiviere, Charles, 36-37
Veraa, Richard, 229
Veterans Administration/Department of Veterans Affairs, xiii, 229
Veterans General Hospital (Taipei), 108
Vidal Sanz, Manuel, 67-68, 69, 72, 232